Lichens above Treeline

Lichens above Treeline

*A Hiker's Guide to Alpine Zone Lichens
of the Northeastern United States*

Ralph Pope

University Press of New England
Hanover and London

Published by University Press of New England,
One Court Street, Lebanon, NH 03766
www.upne.com
© 2005 by University Press of New England
Printed in China
5 4 3 2 1

All rights reserved. No part of this book may be reproduced in any form or by any electronic or mechanical means, including storage and retrieval systems, without permission in writing from the publisher, except by a reviewer, who may quote brief passages in a review. Members of educational institutions and organizations wishing to photocopy any of the work for classroom use, or authors and publishers who would like to obtain permission for any of the material in the work, should contact Permissions, University Press of New England, One Court Street, Lebanon, NH 03766.

ISBN: 1–58465–402–3
Library of Congress Control Number: 2005920241

All photographs are by the author.

FRONT COVER PHOTOGRAPHS: Background image: A view from the Presidentials in the White Mountains, New Hampshire. Top inset photograph: *Melanelia hepatizon*, Mount Katahdin, Maine. Middle inset photograph: *Ophioparma ventosa*, Green Mountains, Vermont. Bottom inset photograph: *Cladonia stygia* above and *Cetraria laevigata* below, White Mountains, New Hampshire.
BACK COVER: *Evernia mesomorpha*.

Contents

Acknowledgments — vii
Introduction — ix
   Lichen Biology — x
   Suggested Reading — xii
   Coverage and Methods — xiii
   Chemical Testing — xiii
   Alpine Zones of the Northeast — xiv
How to Use this Guide — xv
   Don't Forget a Magnifier — xvi
   Master Page Layout — xvii

Special Topics: Lichens as Food — 2
Species Descriptions: Bushy Lichens with Flat Stems — 3
Special Topics: Human Uses of Lichens — 9
Species Descriptions: Bushy Lichens with Round Stems — 11
Special Topics: Lichen Substances — 21
Species Descriptions: Stalked Lichens — 22
Special Topics: Pollution Monitoring with Lichens — 26
Species Descriptions: Foliose Lichens — 28
Special Topics: Lichens and Harsh Environments — 38
Special Topics: Lichens and Radioactivity — 39
Species Descriptions Umbilicate Lichens — 40
Special Topics: Lichens and Wildlife — 46
Special Topics: Crustose Lichen Identification — 47
Species Descriptions: Crustose Lichens — 48

Glossary — 64
References — 66
Species Index — 68

Dedicated to Jean,

for her invaluable assistance with this book,

and for always being there

with support and encouragement.

Acknowledgments

Thanks are due to Tom Wessels, Antioch New England Graduate School, and Donald Pfister, Harvard University, for their encouragement and advice while I was a student at Antioch developing the master's thesis that would eventually become this book. Additional thanks also go to Dr. Pfister, who was more than generous with his time in helping me access the extraordinary resources of the Farlow Herbarium at Harvard University. I would also like to thank several lichenologists much more knowledgeable than I who were of inestimable help as I worked my way through the book-writing process, though I certainly accept the blame for any errors or omissions that survived their efforts to help. Jim Hinds, a Maine lichenologist and alpine zone ecologist, was a great source of help, technical and otherwise, and I am grateful to him for the many hours he devoted to this project. Irwin (Ernie) Brodo, my lichen professor at Eagle Hill Institute, was a great source of inspiration and information as a teacher, a writer, and a reviewer. I am also indebted to Richard Harris, Scott LaGreca, Elizabeth Kneiper, and Nancy Slack for their thoughtful comments on the manuscript.

I would like to thank Rindy Garner and Deb Verhoff for their assistance with graphic design, and Wink Lees, a fellow student at Antioch, for providing invaluable hiking advice and housing in the White Mountains. My Katahdin hiking partners, Tim Vrabel and Richie Estabrook, provided much-needed assistance, advice, and companionship. Dick Fortin and his Antioch Alpine Flora class gave important feedback on the usefulness of the guide in the field.

And finally, many thanks to my wife Jean and our late friend Enfield Ford who, as my tireless editors, hunted down my mistakes and made many valuable suggestions.

Introduction

Here in the Northeast, we enjoy a well-deserved reputation for challenging weather. Storms converging here from the Great Lakes, the Gulf of Mexico, and the Atlantic coast combine with arctic air from the Canadian North to bring us wetter, colder, and stormier weather than our latitude, which is roughly the same as the French Riviera, would suggest. With a unique combination of weather and geography, our mountaintops are pounded by some of the harshest arctic weather in the world (Marchand 1987). The organisms living on these summits all have unique adaptations allowing them to survive in extreme environments. And lichens don't just survive in this punishing environment; they are the dominant life form in much of the alpine zone. This place where only the hardy survive is perfect for lichens, the ultimate colonizers of habitats in harsh environments.

Lichens get this competitive edge from three important adaptations. First, they are cryptobiotic. They have evolved the ability to completely shut down all operations when the going gets tough, and to start up quickly when there's a chance to photosynthesize and make some food (Wessels 2001). Second, they can photosynthesize at temperatures as low as –20C, allowing them to make hay whenever the sun shines (Richardson 1975). And third, except for a few trace elements, they get all the nutrients they need from air, humidity, rain, and snow, thus freeing them to colonize a substrate without regard to nutrient availability (Brodo et al. 2001). Armed with these survival tools, lichens are able to take advantage of the lack of competition from vascular plants, most of which have a difficult time surviving in extremely cold, windswept, or nutrient-poor conditions.

What we refer to as alpine summits are actually remnant arctic ecosystems connecting us ecologically to a time when much of North America was covered by glaciers. As the Northern Hemisphere warmed following the last glaciation and the Laurentide ice sheet began its retreat approximately 14,000 years ago, the arctic tundra ecosystem followed the retreat of the glaciers to the north. As warming continued, the boreal forests gradually replaced the tundra ecosystem and the tundra began a retreat up-slope to ever-smaller patches, eventually occupying their current restricted range on the summits of mountains. They have gone as far as they can go. If the climate continues to warm, this arctic tundra ecosystem will have no place for further retreat. It is literally and figuratively poised at a great precipice as global warming threatens to push our alpine zones over the edge.

In many ways, lichens are perfect subjects for a visual field guide. They occur in all terrestrial ecosystems in the world, they are colorful, and their fourteen thousand species worldwide exhibit astounding variety (Brodo et al. 2001) When you find an interesting lichen to study, you can count on it being where you found it next week or next year—quite unlike its floral cousins, the tough but ephemeral arctic wildflowers that adapt to their short growing season with a rapid transition from flower to seed production. Lichens display their bodies and their reproductive structures throughout the year. The diversity and abundance of lichens provide important information about the health of our alpine zones. My hope is that this field guide will stimulate interest in protecting this land above treeline by helping people understand the critical role that lichens play in this fragile, beautiful, and threatened ecosystem.

Lichen Biology

An understanding of the biology of lichens has been slow in developing. The ancient Greeks and Romans knew lichens well, using them for medicines and to make purple dye, a symbol of wealth and nobility. But they gave them no special place in nature, combining them taxonomically with liverworts and mosses. Not much attention was paid to lichen biology until the eighteenth century, when the revolutionary taxonomist Linnaeus devised the current system for the classification of all living things. Linnaeus clearly understood that they were separate from the mosses, but he accorded lichens only one genus, describing them as "the poor trash of vegetation" (Richardson 1975). Fortunately for us, Edward Acharius, a student of Linnaeus, was greatly excited by lichens and spent his lifetime studying lichens and classifying them into more than forty-one genera, based on the relatively modern concept of classifying organisms according to the morphology of their reproductive structures (Uppsala University 2002).

The next major step in understanding lichen biology was taken in the 1860s, when two European botanists, Simon Schwendener and Anton de Bary, introduced the "dual hypothesis" (Gilbert 2000; Richardson 1975). This hypothesis proposed that the small green cells found in the lichen were not outgrowths of fungal filaments as previously thought, but were photosynthetic algae providing nutrition to the fungus. This concept, so widely accepted today, proved to be highly controversial. Many of the leading lichenologists of the day, including the great American botanist Edward Tuckerman of Tuckerman Ravine fame, never accepted this hypothesis, believing it to be a romantic and even silly concept. And so, as is frequently the case, science marched on one funeral at a time. Today, the much-argued dual hypothesis that lichens are actually two organisms—a fungal partner (mycobiont) and an algal or cyanobacterial (photobiont) partner living in a symbiotic relationship—is now an established fact.

For the past 140 years, theories about the nature of this symbiosis have varied from fungal parasitism, as first proposed by

Schwendener (Purvis 2000), to the currently accepted concept of mutualism where both parties benefit. There isn't any question that the fungus gains. The helpful algae fill all of its energy needs. But in order for this to be mutualistic, the algae must gain also. And gain they do. As a result of their relationship with the mycobiont, the algae get housing, and, they are protected from being eaten, from bacterial infection, from dessiccation, and from competition from plants. And, with their fungal partner, they are able to colonize areas that they could never inhabit successfully on their own. In fact, *Trebouxia*, the most common algal genus in lichens, is rarely found free-living (Brodo et al. 2001; Kendrick 2000). So it's possible that some algae may find survival itself one of the benefits of the association.

    The photosynthetic lichen partner (photobiont) is a green alga about 90 percent of the time and a cyanobacterium 10 percent of the time. Strange as it may seem, two more evolutionarily separated groups of organisms can't be found on the planet than algae and cyanobacteria, representing life forms that diverged approximately 1.7 billion years ago (Campbell et al. 1999). However, each of these photobionts does the job of providing usable energy supplies for their group of lichens. Both types of photobionts combine energy from the sun with carbon dioxide and water to manufacture carbohydrates through photosynthesis. The fungal partner, or mycobiont, pursues a delicate balance, absconding with enough of the photosynthate to survive and grow while leaving just enough nutrition for the photobiont to survive and grow. With a few exceptions, the fungal partner is unique for each lichen, and each fungus will only use the services of a single type, or occasionally two types, of photobiont. However, some species of algae appear to be useful partners to many fungi. Because of this fungal uniqueness, because the fungus produces the conspicuous fruiting bodies, and also because the fungal partner provides about 90 percent of the biomass, the taxonomy—and therefore the scientific names—of lichens are based on the fungal partner. Much remains to be discovered about the photobiont partners. Only about 2 percent of the algae living with lichens have been identified to species, in part because the union substantially changes their morphology. To determine their species, researchers need to follow the difficult course of separating the algae from the fungus and culturing them individually.

    Lichenologists have described approximately fourteen thousand species of lichens worldwide, and undoubtedly many more remain to be discovered (Brodo et al. 2001). The mycobionts are almost all from the Ascomycota group of fungi, or fungi that reproduce sexually by forming spores within structures known generally as ascomata. These structures of sexual reproduction in lichens most frequently take the form of tiny quiche-like shapes called apothecia, and less often take the shape of immersed, vase-shaped structures called perithecia. In a very few lichens, the mycobiont is a member of the Basidiomycota group of fungi that form the familiar mushroom-like structures (basidiomata) that house their spore manufacturing and distributing organs.

Dispersal resulting from sexual reproduction begins when microscopic fungal spores are ejected from ascomata into the air, where they become part of the Aeolian plankton drifting in the wind. In order to form a new lichen, these spores must land, germinate, find, and unite with the right photobiont. The mycobiont can develop its unique lichen morphology only after establishing a relationship with an acceptable photobiont. This is a tall order for three reasons. First, lichens are frequently quite specific about which photobiont they will consort with; second, many of the photobionts are not commonly found free-living in nature; and third, at least in a laboratory, synthesizing a lichen from its component parts is very difficult (Kendrick 2000). Because of these difficulties inherent in sexual reproduction, lichens have developed some very specialized ways of reproducing asexually by distributing parts of themselves that have both the mycobiont and the photobiont pre-packaged and ready to grow. Many lichens accomplish this in one or more of three ways. First, lichens are often very brittle when dry and pieces capable of regenerating a new lichen frequently break off; second, they form isidia, which are very small, easily dislodged hot dog-shaped extensions of the parent lichen that contains both the mycobiont and the photobiont; and third, they produce loose microscopic granules called soredia that are composed of a few algal cells wrapped in threads of fungal tissue. Wind, water, or animals can distribute these fragments, isidia, or soredia, and, given appropriate conditions and substrate, a new lichen begins.

Suggested Reading

The accessibility of good information about lichens changed dramatically in 2001 with the Yale University Press publication of *Lichens of North America* by Irwin Brodo, Sylvia Sharnoff, and Stephen Sharnoff. This is a wonderful reference book for both beginning and experienced lichen enthusiasts. It can be used as an entire course in lichenology, and the identification photos and keying sections are superb. At 8 pounds 9 ounces it's no field guide, but it will be a valued addition to your library.

My other "must have" recommendation is at the other end of the size scale. James and Patricia Hinds have published a very nice *Simplified Field Key to Maine Macrolichens*. It works well throughout the Northeast and is small, light, and user-friendly. There are no illustrations, so once you key out a lichen using this reference you have no idea if you've got it right, but that's why you should have *Lichens of North America* on your desk at home. To order a copy of the Hinds's key, you can reach the authors at jwhplh@earthlink.net.

Coverage and Methods

Fieldwork for this book took place in the alpine zones of the Adirondacks, the Green Mountains, the White Mountain National Forest, and on Mount Katahdin. Pursuant to a research permit graciously provided by the U.S. Forest Service, I photographed and collected voucher samples in the White Mountains for most of the species included in the book. Research in other areas generally was limited to photography and a review of herbarium specimens.

Description, habitat, and range information was compiled from several sources. First, I relied on specimens collected and observations made in the field. Second, I reviewed herbarium specimens at the Farlow Herbarium at Harvard University. And third, I consulted the following references: *American Arctic Lichens, the Macrolichens* (Thomson 1984), *American Arctic Lichens, the Microlichens* (Thomson 1997), *How to Know the Lichens* (Hale 1979), *Lichens of North America* (Brodo et al. 2001), *Macrolichens of the Pacific Northwest* (McCune & Geiser 2000), and the *Simplified Field Key to Maine Macrolichens* (Hinds and Hinds 1992). Whenever discrepancies in descriptions arose, I gave the greatest weight first to the specimens that I collected, and second to Northeastern U.S. specimens on file at the Farlow Herbarium, since these sources provide descriptive information most pertinent to the populations being represented.

All scientific names follow Esslinger's checklist of lichen names (1997) as revised in 2004, and most common names follow Brodo et al. (2001). Scientific name derivations were compiled from information in *Cassell's Latin Dictionary* (Simpson 1968), *A Source-Book of Biological Names and Terms* (Jaeger 1944), *Composition of Scientific Words* (Brown 1956), *Botanical Latin* (Stearn 1992), *Etymologie der Wissenschaftlichen Gattugsnamen der Flechten* (Feige 1998), and the *Dictionary of Root Words and Combining Forms* (Borror 1988).

Pollution sensitivity information generally has not been included. Though lichens are frequently used as pollution monitors (see p. 26), there is no universally accepted measurement of pollution sensitivity. Most studies on the effect of pollutants on lichens have been conducted at elevations well below the alpine zone, so reliable and comparable pollution tolerance information is not available for many of the included species.

Chemical Testing

Some commonly available chemicals often are used in identifying lichens (see p. 21), and they can certainly make identifications more reliable. However, since most hikers don't carry a chemistry kit with them, this book provides identification guidance without reference to the chemical tests that are a traditional part of lichenology. An indepth discussion of chemical testing techniques can be found in *Lichens of North America* by Brodo et al. (2001) (see Suggested Reading above).

Alpine Zones of the Northeast

The mountains and ranges indicated above all have peaks higher than 4,000 feet (small triangles) to 4,500 feet (large triangles), and generally can be expected to have extensive alpine zones. In addition, there are many lower-elevation summits throughout the Northeast where a hiker can find vibrant alpine plant communities and most of the lichen species in this book.

Northern Presidentials from Mount Washington.

How to Use this Guide

Please take a few minutes to review the method behind the organization of this field guide and become familiar with the page layout shown on page xvii. The four identification steps are:

1: Identify the lichen growth form among the following choices and look for the icon representing that growth form in the top corner of the species description pages.

Bushy, shrubby lichens with stems that are flat to curled but not round and are generally tangled. If these stems are very curled, an open seam still will be visible. Most of them grow in clumps on soil or with moss. Pages 3–8.

Bushy, shrubby lichens with round stems that are interlocked and generally tangled. Even if not perfectly round, these stems will not have an obvious seam. They grow in clumps on the ground or sometimes on rock. Pages 11–20.

Stalked lichens have round hollow stems that stand individually and usually do not grow in clumps. The tops may have cups, points, branches, or globular fruiting bodies, and the bases frequently are surrounded by small leafy growths (squamules) that give rise to the stalks (podetia). Pages 22–25.

Foliose lichens have a distinct upper and lower surface, lobed margins, and tend to grow outward in all directions. They grow on bark, soil, wood, or rock, and can be separated from the substrate, though sometimes with difficulty. Pages 28–37.

Umbilicate lichens also have a distinct upper and lower surface and a tendency to grow outward in all directions. They generally are not lobed, they look like brown or black potato chips, and they grow attached to rocks by a central holdfast (umbilicus). Pages 40–45.

Crustose lichens grow flat against a surface. They have no lower cortex, are intimately attached to their substrate, and cannot be separated from it. They grow on soil, moss, dead vegetation, rock, bark, or wood. Please read the special crustose identification notes on page 47. Pages 48–63.

2: Choose from the following substrate choices within the appropriate growth form section. On the species pages, the substrate icon is immediately below the growth form icon.

 On the ground. Includes anything from bare soil to moss.

 On rock.

 On krummholz trees. Krummholz is a German word meaning "crooked wood," referring to trees in the alpine zone exhibiting a predominately horizontal growth form.

3: Look for the color that best describes your sample. The species pages are arranged by color from lightest to darkest within substrate categories. When light to dark isn't clear, the brightest colors such as yellow or orange are before the grays. In many species, shade-grown specimens will be light, and specimens grown in full sun will be darker, so be sure to check all species pages in a section if you don't find a good match right away. Also, wet lichens are brighter and frequently greener than dry lichens. Most of the photographs in this book are of samples that were at least slightly damp.

4: Review the images and the descriptive information to confirm a correct identification. If there are confusing choices, be sure to read the "Similar Species" section for each possibility. The smaller, circular images represent the view through a ten- to twenty-power hand lens unless indicated differently. As in any other scientific field, lichenology has its own set of specialized terminology, so check the glossary on page 64 if you're not sure of the meaning of any terms. Please note the ruler on the inside back cover.

Don't Forget a Magnifier

A hand lens or jeweler's loupe is a critical tool used to identify these frequently small dual organisms. Each species page has an overall habitat photo that will help you recognize the general growth form, but in most instances a closer look will be required. A ten-power to twenty-power hand lens will enable you to see the detail shown on most of the species pages. Don't leave home without it.

Hand lenses can be purchased from scientific supply houses, most university bookstores, stamp and coin stores, and photographic equipment stores. Or, if you have an old, out-of-alignment pair of binoculars, you can remove the front lens elements and mount them together to make a great hand lens and put those useless old binoculars back to work. In many pairs of binoculars, the eyepieces used individually make a good, if somewhat powerful, hand lens.

Master Page Layout

Scientific Name   Common Name   growth form

substrate

A color bar here shows the typical color or range of colors.

photo details

habitat photo

25 mm scale bar (1 inch)

Descriptive notes here

Many pages include a view such as you might see through your hand lens. → Detail photo at 10X to 20X magnification

Species names in bold type have a description page page in this book.

SIMILAR SPECIES: This section will focus on differences between easily confused species.

RANGE: This section provides information about the geographical distribution of a species.

REPRODUCTION: Describes the reproductive structures present.

NAME: Derivation of the scientific name. (L) = Latin, (G) = Greek.

SPECIAL TOPICS: Gives the page numbers of pertinent essays.

Lichens above Treeline

Lichens as Food

If you're planning to survive in the wilderness on a diet of lichens, you'd better think again. With a few interesting exceptions, they taste bad, they're tough, and they're not very digestible. However, arctic explorers have eaten umbilicate lichens (p. 40–45), also known as "rock tripe," as emergency survival food; and while they may have saved some from starvation, it is questionable whether the caloric gain was worth the diarrhea and nausea (Pielou 1994). As a matter of fact, lichens are so disagreeable that starving explorers ate their boots first, then the lichens (Merrill 1915). The problem seems to be that the complex carbohydrates that form much of the lichen are difficult to digest, and the lichen substances (see Lichen Substances, p. 21) that constitute as much as 5 percent of the lichen mass, commonly have a high acid content (Burt 2000). Boiling in several changes of water with the addition of baking soda or ashes would have made the rock tripe more edible, but baking soda and firewood are generally not available to starving arctic explorers (Burt 2000).

Some lichens are not quite as difficult to digest as the umbilicates. Horsehair lichen (*Bryoria* spp.) is referred to as pine moss by aboriginal people of Canada and is cooked down to a gelatinous black paste that can be shaped into edible biscuits (Brodo et al. 2001; Richardson 1975). Iceland lichens (p. 5, 6) are another reasonably edible lichen group. They can be boiled with an alkali such as baking soda or ash, then dried, powdered, and mixed with flour to make bread or added to soup as a thickener (Richardson 1975).

A Middle Eastern lichen, *Aspicilia esculenta* (also known as *Lecanora esculenta)*, can be ground and mixed with traditional meal to make bread. *Aspicilia esculenta* grows in the deserts of the Middle East, naturally breaks up into chunks approximately 10 millimeters diameter by 6 millimeters thick, and is easily dislodged from its substrate. Apparently these lichen pieces can break loose in a windstorm and collect in protected areas. Some think that this may have been the "manna from heaven" eaten by the Israelites during their exile from Egypt (Merrill 1915; Richardson 1975).

Caribou have symbiotic bacteria in their rumens that help break down the lichen carbohydrates, enabling them to access the calories without the difficulties that humans encounter (Marles et al. 2000). The Inuit people consider partially digested lichen in the rumen of caribou to be a great delicacy, and they eat the mixture right out of the stomach of the freshly killed caribou (Brodo et al. 2001; Marles et al. 2000; Richardson 1975). This "caribou ice cream" is reported to be tastier in the winter when the stomach contents are almost completely lichens than in the summer when many plants are being digested with the lichens (Marles et al. 2000). Could Caribou Crunch be the next Ben and Jerry's hit?

Flavocetraria nivalis — Snow lichen

East slope of Mount Washington, White Mountains, N.H.

25 mm

pale greenish yellow to straw

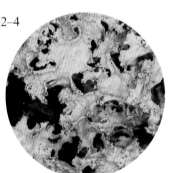

This species forms low-growing tufts (2–4 cm high) in arctic-alpine tundra heath communities. As shown on the detail photo, the thallus lobes are flat (3–6 mm across) with small black dots near the edges and a wrinkled surface. The very divided lobe edges give the appearance of fine ruffles.

SIMILAR SPECIES: The lobes of *F. nivalis* are more wrinkled than *F. cucullata* (next page), and the small black dots are closer to the edges, but it generally looks like an *F. cucullata* that didn't curl up.

RANGE: Circumpolar arctic and boreal following higher terrain south to New England and the eastern Adirondacks.

REPRODUCTION: Apothecia are rare, and there are no soredia or isidia. Reproduction is probably by fragmentation.

NAME: *Flavidus* (L) = yellow, *cetraria* (L) = a genus name meaning like a shield, *nivalis* (L) = pertaining to snow; that is, a yellowish lichen related to the genus *Cetraria*, growing in arctic-alpine (snowy) places.

Flavocetraria cucullata — Curled snow lichen

pale greenish yellow to straw

East slope of Mount Washington, White Mountains, N.H.

The branches of this lichen grow fairly vertically, forming loose to tight tufts 25 to 50 mm tall. Typical branches are less than 5 mm across with edges rolled inward to form incomplete tubes. Branches are smooth to slightly pitted with tips that are ruffled and bend backwards. This species prefers moss or organic soil in late snow-melt areas.

SIMILAR SPECIES: *F. nivalis* (previous page) is very similar, but its branch edges are inrolled only slightly if at all.

RANGE: Circumpolar arctic and boreal following higher terrain south to New England and the eastern Adirondacks.

REPRODUCTION: Apothecia are rare, and are usually found only on dense tufts in late snow-melt areas. There are no soredia or isidia.

NAME: *Flavidus* (L) = yellow, *cetraria* (L) = a genus name meaning like a shield, *cucullata* (L) = curled; that is, a yellowish lichen with curled branches related to the genus *Cetraria*.

Cetraria laevigata   **Striped Iceland lichen**

yellow-green to brown

Gulfside Trail on Mount Clay, White Mountains, N.H.

The branches are usually under 5 mm wide with marginal pseudocyphellae (white lines, see arrow) frequently continuous from tip to base. The smooth-surfaced branches are inrolled almost, but not quite, forming tubes, with tips that usually bend backwards. Marginal projections are blunt-tipped and occasionally branched.

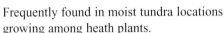

Frequently found in moist tundra locations growing among heath plants.

SIMILAR SPECIES: The very similar *C. islandica* (not illustrated) has white pseudocyphellae on the surface, less continuous marginal pseudocyphellae, and is rare in our alpine areas.

RANGE: Widespread throughout Alaska and arctic Canada, south into the alpine zones of northeastern United States.

REPRODUCTION: There are no soredia, or isidia and apothecia are infrequent. Most reproduction is probably by fragmentation.

NAME: *Cetraria* (L) = like a shield, *laevigata* (L) = smooth; that is, a lichen in the genus *Cetraria* with smooth branches.

SPECIAL TOPICS: Lichens as Food, p. 2; Human Uses of Lichens, p. 9.

Cetrariella delisei Snow-bed Iceland lichen

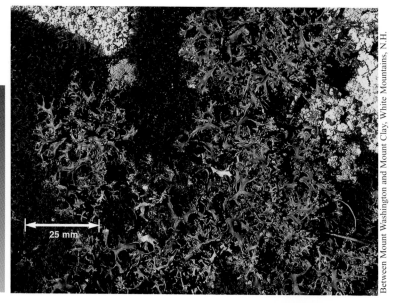

Between Mount Washington and Mount Clay, White Mountains, N.H.

yellow-brown to dark brown

This very finely dissected cousin of the much larger **Cetraria laevigata** (p. 5) doesn't look like much more than a brown tangle until you get out your hand lens and look at the structure in detail. The thallus is composed of straps less than 1 mm across tending to curl slightly at the edges but remaining mostly flat and terminating in fine lacy-looking fingers (see detail photo). This alpine lichen is most common in moist snowbank communities that have snow cover well into the alpine summer.

SIMILAR SPECIES: Both **Cetraria laevigata** and *C. islandica* are significantly larger with marginal projections and more pronounced edge roll.

RANGE: Circumglobal arctic and alpine in Europe and eastern U.S.

REPRODUCTION: There are no apothecia, soredia, or isidia. Reproduction is probably by fragmentation.

NAME: *Cetrariella* (L) = like a small shield, *Delise* = a French botanist; that is, a small lichen similar to lichens of the genus *Cetraria*, named for the botanist D. F. Delise (1780–1841).

SPECIAL TOPICS: Lichens as Food, p. 2; Human Uses of Lichens, p. 9.

Cladonia strepsilis　　　　　　　　Ball lichen

South Baldface Mountain, on granite, N.H.

25 mm

yellow-green

The entire thallus is composed of a tight cluster of very fine squamules curled up into a hollow ball that can be as small as 25 mm across, like the sample in the photograph above, or as large as 50 mm across. There may be a few very small podetia not much higher than the squamules with brown apothecia at the tips. It is common on granite and on soil in open areas between boulders. When growing on soil, it doesn't form balls and is more likely to form podetia.

SIMILAR SPECIES: Many *Cladonia* species begin life as a mat of squamules and only develop their distinctive podetia as they mature, so many squamulose lichens similar to this example will be impossible to identify in the field. The balled-up growth form and scattered small brown apothecia not much higher than the edge of the ball should be helpful.

RANGE: Not an arctic lichen, this species ranges in North America from the Canadian Maritimes to the U.S. Gulf States.

REPRODUCTION: Small brown apothecia occasionally are produced on short podetia.

NAME: *Cladonia* (G) = with branches or stalks, *strepsilis* (G) = twisted; that is, a curled up clump of stalks.

Evernia mesomorpha — Boreal oakmoss lichen

yellow-green

Mount Moosilauke on the Carriage Road Trail, White Mountains, N.H.

25 mm

This common subalpine lichen forms occasionally pendent tufts approximately 4 to 8 cm high with branches that are irregularly flattened, wrinkled, and usually covered with scattered patches of granular soredia on both the flat surfaces and the edges, as shown in the inset. It's included here with the alpine species because it occasionally grows in small tufts on krummholz bark well above the treeline. Just below the alpine zone where trees are short, this sun-loving lichen is very common.

SIMILAR SPECIES: *Usnea* lichens have round branches with a central cord. *Ramalina* lichens have stiff branches with soredia in marginal or apical soralia rather than the flexible branches and diffuse soredia in *Evernia*. Neither *Usnea* nor *Ramalina* lichens are alpine.

RANGE: Canadian arctic temperate and boreal but not extending to the Pacific. In the East, it is found south to West Virginia.

REPRODUCTION: Soredia and fragmentation. Apothecia unknown.

NAME: *Evernes* (G) = branching, *mesomorpha* (G) = intermediate shape; that is, a branched lichen with a form intermediate between other species.

SPECIAL TOPICS: Human Uses of Lichens, p. 9.

Human Uses of Lichens

Many uses for lichens have been recorded over the centuries, with medicines and dyes leading the way. During the Middle Ages, the medicinal uses of plants were largely governed by the "Doctrine of Signatures," which suggests that God made certain plants or plant parts to look like the diseased body parts or organs they were meant to heal. Following this rather bizarre theory, *Lobaria pulmonaria* (lungwort) was used to treat diseases of the lungs because of its resemblance to lung tissue, and the hair-like lichens of the genus *Usnea* were used to treat diseases of the scalp. An equally strange practice in the Middle Ages was to use any lichen growing on human skulls to treat epilepsy and other serious diseases (Purvis 2000). The herb garden must have been quite a sight. In most early medical uses, a powdered lichen was administered to the patient in an alcoholic mixture, which probably helped lessen the patient's suffering even if any real curative properties were lacking.

Fortunately, some medicinal uses have a more solid scientific basis. *Usnea* species have been used for centuries in medicines, and modern testing has shown that a chemical found in usnea, usnic acid, does indeed have antibiotic properties (Brodo et al. 2001). Tom's of Maine uses an extract from *Usnea* species as an ingredient in their underarm deodorant. And for probably unrelated but little understood reasons, these useful *Usnea* species have been found between the wrappings of Egyptian mummies (Merrill 1915). Iceland lichens contain substances with antibiotic properties and were used as a treatment for tuberculosis and other respiratory ailments before the development of modern drugs (Richardson 1975). The versatile Iceland lichens have also been used as food and as a heat-stable antibiotic food additive to retard spoilage (Richardson 1975), as a food thickener, and as a treatment for athlete's foot and ringworm (Marles et al. 2000). The visually delightful pixie cups (*Cladonia* spp.) have been used in Europe as a remedy for coughs (Merrill 1915).

Some of the Dene tribes of Canada use a syrup made by boiling rock tripe (*Umbilicaria* spp.) to expel tapeworms (Marles et al. 2000), which may explain why this lichen functions so poorly as an emergency food (see Lichens as Food, p. 2). The Dene also use ***Evernia mesomorpha*** (p. 8) and *Usnea hirta*, two lichens with usnic acid, to treat snow blindness and to stop nosebleeds.

Roccella tinctoria, a Mediterranean lichen, was used by ancient Egyptians, Phoenicians, Greeks, and Romans to make purple dye, a highly valued color in the ancient and medieval worlds. This same lichen was the original source of blue dye for litmus paper. The original Harris tweeds from the Harris Islands of Scotland were colored with lichen dyes, a practice that continued in parts of the islands until the 1970s (Richardson 1975).

Model railroad buffs will recognize many of the *Cladonia* genus alpine zone lichens as the miniature shrubs and trees sold to make train sets look realistic. These same groups of lichens are also used in architectural modeling and in floral decorations (Marles et al. 2000).

Because usnic acid inhibits the manufacture of photosynthetic pigments and can reduce photosynthesis in plants, lichens are currently being investigated as possible environmentally friendly herbicides. In addition, many other of the over one thousand chemicals produced by lichens (see Lichen Substances, p. 21) are being investigated for possible medicinal uses.

Two closely related reindeer lichens on Mount Katahdin, Baxter State Park, Maine. The slightly darker population roughly above and left of the line is **Cladonia stygia** and the lighter population in the bottom right is *C. rangiferina*. See page 13.

Stereocaulon alpinum   Earth foam lichen

Mount Katahdin, late snowmelt area on the Hamlin Ridge Trail, Maine

white

There are a number of *Stereocaulon* species in the alpine zone that can be difficult to distinguish. They are all light gray to white, most grow on rocks (but this species grows on moss or soil), they like full sun, and they have a low growth form that frequently looks like fine seafoam. Their solid stalks are covered with lobules referred to as phyllocladia. ***Stereocaulon alpinum*** grows in colonies 5 to 10 cm across and 1 to 3 cm high with stout stalks that are almost completely covered with very fine, ocasionally branched, granular phyllocladia.

SIMILAR SPECIES: ***Stereocaulon alpinum*** is the only *Stereocaulon* in our alpine zones not found on rock. See pages 19 and 20.

RANGE: Circumpolar arctic-alpine, south to New Hampshire in the East. Also in the mountains of western North America and in South America.

REPRODUCTION: No soredia or isidia. Brown apothecia are occasionally produced at the ends of the branches, though there are none on the sample above.

NAME: *Stereo* (G) = solid, *caulon* (G) = stalk, *alpinum* (L) = alpine; that is, alpine and solid stalked.

Thamnolia vermicularis   Whiteworm lichen

white

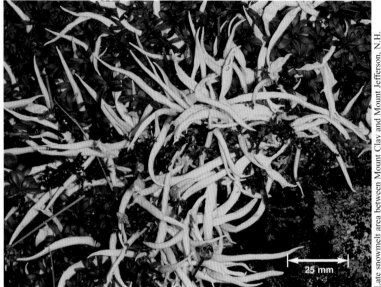

Late snowmelt area between Mount Clay and Mount Jefferson, N.H.

This distinctive lichen is composed of unbranched to slightly branched, round, bone-white, hollow podetia 2 to 4 cm long with points at both ends. A substrate generalist, **T. vermicularis** will grow mixed with mosses and heath plants or by itself on exposed gravelly soil. In slight chemical variants, it is found throughout the world in such diverse locations as New Guinea, New Zealand, Australia, and the Andes Mountains of South America, as well as in arctic and alpine areas of North America. This broad dispersal is quite surprising in light of the lack of reproductive structures. Noting the absence of this species on active or resting volcanoes where it might have been destroyed, Thomson (1984) suggests that this distribution may be ancient, predating the separation of the continents as we know them today.

SIMILAR SPECIES: There are no similar species.

RANGE: Circumpolar arctic-alpine with worldwide distribution.

REPRODUCTION: This species produces no apothecia, soredia, or isidia. Reproduction is presumably by fragmentation.

NAME: *Thamnolia* (G) = bush or shrub-like, *vermicularis* (L) = worm; that is, a bushy (?) worm-shaped lichen.

SPECIAL TOPICS: Lichens and Wildlife, page 46.

Cladonia stygia   Black-footed reindeer lichen
Formerly *Cladina stygia*

This very common shrubby lichen frequently forms bushy clumps as thick as 10 cm. The smooth, cotton-surfaced stalks are round, branched, and tangled with usually browned tips typically all bending in the same direction. The bases of the stalks are black inside, and black with white splotches on the outside. Stalks are more scattered and less shrubby at higher elevations. This species, with *C. rangiferina*, **C. stellaris**, **C. arbuscula**, and **C. uncialis**, forms the major part of a caribou's winter diet.

SIMILAR SPECIES: The only similar gray lichen is *C. rangiferina,* which also can be alpine and looks very much the same but doesn't have the blackened base. See page 10 for comparison. ***Cladonia arbuscula*** is also similar in shape and style but is yellowish green (p. 15).

RANGE: Circumpolar arctic to boreal, extending into the Northeast at high elevation. *C. rangiferina* is more common at lower elevations.

REPRODUCTION: Apothecia are rare, and no soredia or isidia are produced. Reproduction is presumably by fragmentation.

NAME: *Cladonia* (G) = with branches or stalks, *stygius* = dark as the underworld river Styx; that is, referring to the blackened base.

SPECIAL TOPICS: Lichens and Wildlife, page 46.

Cladonia stellaris   Star-tipped reindeer lichen
Formerly *Cladina stellaris*

pale yellow-green to gray

Between Mount Clay and Mount Jefferson, White Mountains, N.H.

25 mm

The round thallus branches form tight, tufted heads approximately 2 to 4 cm across and 6 to 10 cm tall. Sometimes these heads form extensive mats on soil, mossy humus, or among heath plants. The podetia branch into star-like clusters of three to six tips, giving this lichen its species name. ***Cladonia stellaris*** has many uses. Caribou prefer this over all other reindeer lichens; it is the lichen most often used for model railroad shrubs and trees; and it is used in Europe as a source of usnic acid, a mild antibiotic (Richardson 1975).

SIMILAR SPECIES: The light green to whitish color and tight compact "cotton ball" heads are very distinctive.

RANGE: Circumpolar arctic and boreal, extending south to West Virginia in the higher terrain of the Appalachians.

REPRODUCTION: Apothecia are very rare, and no soredia or isidia are produced. Reproduction is presumably by fragmentation.

NAME: *Cladonia* (G) = with branches or stalks, *stellaris* (L) = starry; that is, a branched lichen with star-shaped tips.

SPECIAL TOPICS: Lichens and Wildlife, page 46.

Cladonia arbuscula   Reindeer lichen

Formerly *Cladina arbuscula*

Dudley Trail, Mount Katahdin, Baxter State Park, Maine.

yellow-green

This bushy lichen looks very much like a yellow-green version of ***C. stygia*** (p. 13). It forms tufts 5 to 10 cm high by 10 cm or more wide with slightly browned tips that frequently have a combed look. The round, hollow main branches are less than 2 mm across, with a green-splotched outer surface.

SIMILAR SPECIES: ***Cladonia arbuscula*** can be distinguished from ***C. stygia*** and *C. rangiferina* (see p. 10) by its yellow-green color. The more delicate, combed terminal branches separate this from ***C. uncialis***. It looks very much like *C. mitis* (not shown), which is probably not alpine, is chemically different, and is less apt to have the combed look.

RANGE: Circumpolar arctic and boreal forest lichen extending south in the East into the Appalachians.

REPRODUCTION: Apothecia are rare and no soredia or isidia are produced. Reproduction is probably by fragmentation.

NAME: *Cladonia* (G) = with branches or stalks, *arbuscula* (L) = a small tree; that is, a small tree or bush.

SPECIAL TOPICS: Lichens and Wildlife, page 46.

Cladonia uncialis — Thorn cladonia

yellow-green

Gray Knob Trail, White Mountains, N.H.

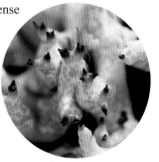

The abundantly branched podetia form dense tufts 2 to 8 cm tall and as large as 20 cm across but usually much smaller. The terminal branches are frequently browned at the tips, pointed, and sometimes are arranged in a whorl, but they do not form cups. The fine podetia are 1 to 1.5 mm in diameter with a splotchy yellow-green surface, and are very fragile when dry.

SIMILAR SPECIES: ***Cladonia amaurocraea*** is more delicate looking, has longer tips, and frequently has at least a few cups. ***Cladonia arbuscula*** has branch tips that are much more delicate and combed, and the branch surface of ***C. arbuscula*** is cottony, while the surface of ***C. uncialis*** is shiny.

RANGE: Widespread circumpolar arctic, boreal, and temperate, extending south to Georgia.

REPRODUCTION: No apothecia, soredia, or isidia are produced. Reproduction is probably by fragmentation.

NAME: *Cladonia* (G) = with branches or stalks, *uncinatus* (L) = barbed; that is, a branched lichen with barb-like tips.

SPECIAL TOPICS: Lichens and Wildlife, page 46.

Cladonia amaurocraea — Quill lichen

yellow-green to brown

The thallus is composed of round, shiny, usually yellow-green branches forming tufts 6 to 10 cm tall and 4 to 10 cm across. The main branches are usually low to the ground with tips that frequently point upwards. Subalpine populations will usually have small, rapidly flaring cups with small marginal projections as shown in the top detail photo, while many of the alpine examples will have fewer cups and more pointed tips as shown in the lower detail photo.

SIMILAR SPECIES: ***Cladonia uncialis*** is more branched, and has shorter, more blunt tips. Some forms of *C. gracilis* are very similar, but are gray-green, rather than yellow-green to brown, don't have cups, and are less apt to be alpine.

RANGE: Circumpolar arctic and boreal, south to the alpine zones of the Northeast.

REPRODUCTION: No soredia or isidia are produced, and apothecia are rare. Reproduction is probably by fragmentation.

NAME: *Cladonia* (G) = with branches or stalks, *amaur* (G) = dark, *ocrea* (L) = sheath; that is, with dark stalks.

Sphaerophorus fragilis — Fragile coral lichen

red-brown

25 mm

Near the Mount Washington Auto Road, White Mountains, N.H.

The solid, coral-like cylindrical branches form very dense low tufts approximately 2 to 3 mm tall and 3 to 6 cm across on soil, rock, moss, or scat. The color can range from mottled brown in the full sun to quite light in shade. Note the scale bar above. From a distance, this very finely divided lichen looks almost like a dense fur.

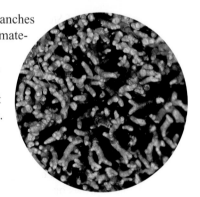

SIMILAR SPECIES: None.

RANGE: This is an arctic tundra species that appears in the alpine areas of the Northeast.

REPRODUCTION: No soredia or isidia are produced, apothecia are rare, and most reproduction is probably by fragmentation.

NAME: *Sphaerophorus* (G) = bearing spheres, *fragilis* (L) = brittle; that is, a brittle lichen with round fruiting bodies (commonly seen on another species in this genus).

Stereocaulon glaucescens   Seafoam lichen

There are a number of *Stereocaulon* species in the alpine zone that can be difficult to distinguish. They are all light gray to white, many grow on rocks, they like full sun, and they have a low growth form that frequently looks like fine seafoam. Their stalks are covered with lobules referred to as phyllocladia. ***Stereocaulon glaucescens*** grows on rock in colonies 5-10 cm across and 1-3 cm high with stout stalks that are almost completely covered with flattened phyllocladia.

SIMILAR SPECIES: ***Stereocaulon alpinum*** (p. 11), also common in the White Mountains and on Katahdin, is very similar but has finer phyllocladia, it grows on soil or with moss, and it has a more upright growth pattern. ***Stereocaulon dactophyllum*** on the next page has cylindrical, branched phyllocladia and brown apothecia are common. *Stereocaulon saxatile* (not illustrated) resembles ***S. glaucescens*** but usually has a fine gray tomentum, a flatter growth pattern and is more likely to be found below treeline.

RANGE: Arctic Canada, the Great Lakes, and northeastern US.

REPRODUCTION: No soredia or isidia are produced and there are usually no apothecia. Reproduction is presumably by fragmentation.

NAME: *Stereo* = solid (G), *caulon* = stalk (G), *glaucescens* = becoming bluish gray (G); that is, gray and solid stalked.

Stereocaulon dactylophyllum
Finger-scale foam lichen

Crawford Trail south of Mount Washington, White Mountains, N.H.

25 mm

The specimen pictured is a particularly fertile sample of **S. dactylophyllum**, but the brown globular apothecia at the ends of the stalks are usually quite common. The phylocladia that cover the stalks are elongated and somewhat finger-shaped as indicated by the scientific name. Its preferred habitat is siliceous rock.

Because many *Stereocaulon* lichens are tolerant of heavy metals, they can be analyzed to monitor atmospheric heavy metal deposition (Richardson 1992).

SIMILAR SPECIES: See **S. glaucescens**.

RANGE: Northeastern United States and Maritime Canada.

REPRODUCTION: Globular brown apothecia up to 2 mm across are common.

NAME: *Stereo* (G) = solid, *caulon* (G) = stalk, *dactylo* (G) = finger, *phyllum* (G) = leaf; that is, solid, stalked lichens with finger-shaped phyllocladia.

SPECIAL TOPICS: Pollution Monitoring with Lichens, page 26.

Lichen Substances

The oldest lichen fossils found to date are over 400 million years old, making lichens more than twice as old as the flowering plants that now dominate vegetation on earth (Purvis 2000). Lichens have used their time wisely, evolving over a thousand diverse chemicals, most of which are at least mildly acidic and are referred to as lichen acids. Most of these lichen acids are stored in crystal form on the surface of the fungal hyphae, and though the function of many of them is not understood, the services they provide to the lichen seem to fall into the following categories:

Gas exchange. Collecting on the surface of the fungal hyphae, crystals repel water and provide air spaces facilitating the gas exchange required for photosynthesis (Brodo et al. 2001).

Protection from herbivory. Some of the more foul-tasting lichen acids are concentrated in the cortex and are believed to help keep the lichen from being eaten by herbivores (Gilbert 2000). See Lichens as Food, page 2; Lichens and Wildlife, page 46; and Lichens and Radioactivity, page 39.

Protection from ultraviolet light. Studies have demonstrated that pigments common in high-altitude lichens provide protection from harmful UV rays that could damage the photobiont (Gilbert 2000).

Protection from bacteria. Some of the substances have demonstrated an ability to protect against incursions by bacteria, fungi (Richardson 1975), and other lichens (Human Uses, page 9).

Allelopathy. This is the characteristic of being able to limit competition from other organisms by producing a compound that inhibits their growth. Some lichen compounds that may be able to inhibit the growth of higher plants are being researched (Fahselt 1996).

A lichen's acids provide a unique fingerprint that can be a great help in identifying the species. Three chemicals react with a lichen's chemistry and commonly are used in lichen identification. They are applied to a small spot on a specific part of the lichen and color changes are noted. The three common spot test chemicals are: (1) Potasssium hydroxide solution, abbreviated KOH, or just K. It isn't readily available but is closely related to Draino, which can be used as a substitute; (2) Sodium hypochlorite, abbreviated C, is common household bleach; and (3) *Para*-phenylenediamine, abbreviated PD, or just P, is a nasty carcinogen well worth avoiding. Iodine and ultraviolet light also are used to read certain chemical signatures. Most of the species shown in this book can be identified without chemicals.

Most species of lichen have a diagnostic, nonvariable chemical signature. Occasionally, however, species will occur with two or more slightly different chemical signatures called chemotypes. An example in this book is ***Cladonia squamosa*** (p. 25), which exists as one chemotype with squamatic acid and another less common chemotype with thamnolic acid.

Cladonia sulphurina — Greater sulphur-cup

greenish yellow

Gray Knob Trail near The Quay, White Mountains, N.H.

These stalks (podetia) are tall, up to 8 cm high, hollow, with frequent splits, a generally shredded appearance, and a covering of very fine, powdery soredia. The primary squamules around the base of the podetia are sometimes sorediate and may be either small, like the ones shown, or large. Small red apothecia may be present at the tips of the podetia. This species requires a high humus content substrate; it is common on rotting logs or humus-rich soil in lowland areas, in krummholz, and in outcrop or talus areas.

SIMILAR SPECIES: *Cladonia deformis* is similar, but with fewer podetial lacerations, more regular cups, and primary squamules lacking soredia. ***Cladonia pleurota*** has shorter, vase-shaped cups without splits, and granular soredia

RANGE: Circumpolar arctic and boreal, south into the alpine zones of the Northeast.

REPRODUCTION: Small red apothecia are occasionally present, and fine soredia are on both the primary squamules and podetia.

NAME: *Cladonia* (G) = with branches or stalks, *sulphurina* (G) = yellow; that is, a yellowish, stalked lichen.

Cladonia pleurota — Red-fruited pixie-cup

Franconia Ridge, White Mountains, N.H.

yellowish-green

This species has short (up to 2 cm high) yellow-green podetia with granular soredia on the outside and in the cups. The cup edges are even and frequently have small red apothecia on the edge. Primary squamules (squamules at the base of the podetia) can be small or large and are usually deeply lobed. The preferred substrate is soil or rotting wood or bark.

SIMILAR SPECIES: Specimens with well-developed apothecia can resemble *C. coccifera*, but *C. pleurota* podetia have granular soredia and *C. coccifera* podetia are areolate to squamulose (see p. 24).

RANGE: Circumpolar arctic, boreal, and temperate, south to the Gulf States.

REPRODUCTION: This species produces bright red apothecia and granular soredia.

NAME: *Cladonia* (G) = with branches or stalks, *pleurota* (G) = on the side; that is, possibly referring to granular soredia on the sides of the stalks.

Cladonia coccifera — Cornucopia lichen

yellow-green to light brown

25 mm

Gray Knob Trail near The Quay, White Mountains, N.H.

A cup-forming lichen with podetia 10 to 20 mm high, cups 4 to 10 mm across, and large red apothecia on the cup edges. The accompanying photos show the somewhat ragged-looking lobules on the podetia surface and inside the cups. Its preferred substrate is soil and humus in arctic-alpine areas.

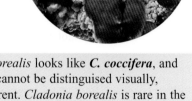

SIMILAR SPECIES: *Cladonia borealis* looks like ***C. coccifera***, and where their ranges overlap they cannot be distinguised visually, though they are chemically different. *Cladonia borealis* is rare in the Northeast, and has not been reported from our alpine regions.

RANGE: Circumpolar arctic and boreal, south in the East to New York State.

REPRODUCTION: Large red apothecia are common, but no soredia or isidia are produced.

NAME: *Cladonia* (G) = with branches or stalks, *coccifera* (L) = bearing scarlet berries; that is, a stalked lichen with scarlet, berry-like apothecia.

Cladonia squamosa — Dragon cladonia

Crawford Path, near Mount Eisenhower, White Mountains, N.H.

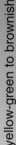

yellow-green to brownish

The basal squamules are finely divided, and the podetia are covered with squamules. This is a highly variable species with podetia up to 5 cm tall, with or without cups. As with many lichens, shade-grown specimens will be light while specimens grown in full sun will be darker. Sometimes found on open sites, but more commonly on rotting wood or tree bases in moist areas. Occasional, but not common in the alpine zone.

SIMILAR SPECIES: This species' finely divided basal squamules and squamulose podetia frequently with brown (never red) apothecia should differentiate it from most similar species. To confirm, a sample under ultraviolet light should fluoresce ice blue.

RANGE: Circumpolar arctic to temperate extending south to Florida.

REPRODUCTION: Brown apothecia frequently are produced.

NAME: *Cladonia* (G) = with branches or stalks, *squamosa* (L) = covered with scales; that is, a lichen with scaly stalks.

SPECIAL TOPICS: Lichen Substances, page 21.

Pollution Monitoring with Lichens

It has long been known that lichens are sensitive to airborne pollution. In 1790, Erasmus Darwin, grandfather of Charles Darwin, and well-known free-thinking poet and scientist, bemoaned the loss of lichens in an area of copper mining and smelting in the following poem from his book *The Botanic Garden* (1790):

> No grassy mantle hides the sable hills,
> No flowery chaplet crowns the trickling rills,
> Nor tufted moss nor leathery lichen creeps,
> In russet tapestry o'er the crumbling seeps.

The first to look scientifically at the effects of atmospheric pollution on lichens was William Nylander, a Finnish botanist working in Paris, who published studies in 1866 and 1896 detailing the decline of lichens in the gardens of Paris (Nylander 1896). However, the level of attention paid to lichens as pollution monitors remained low until the 1950s, when British lichenologists, appalled by the high level of air pollution in urban areas of England, began documenting the tolerance limits of various lichen species. In the 1960s, Brodo published a study of the effects of pollution on the lichen flora of Long Island (1966), and in 1970 Hawksworth and Rose published the first scale for estimating the amount of sulphur dioxide pollution by examining the lichen flora of an area. These two seminal works spawned hundreds of pollution monitoring studies. Currently over 100 (123 in 1999) scientific articles are published each year dealing with lichens and pollution (Henderson 2000).

Lichens have several attributes that make them excellent for pollution monitoring:

Sensitivity to airborne chemicals. Lichens are particularly sensitive to airborne chemicals. They have no stomata and no cuticle to control access to the interior of the thallus, so whenever moisture is present in either liquid or vapor form, dissolved atmospheric nutrients can be absorbed efficiently by the lichen (see Lichens and Radioactivity, page 39).

Bioaccumulators of pollution. Once absorbed, lichens have no mechanism to rid themselves of toxic substances (Ferry et al. 1973).

Variability among species. There is great variability in pollution sensitivity among species. Generally, the fruticose species with their very high surface to volume ratio are the most sensitive, and crustose species the least, but there is much variation within these growth forms. Species with cyanobacteria are significantly more sensitive to pollution than species with algae as the photobiont (Gilbert 2000).

There is a range of reaction within a species. Travelling from an area of low pollution to an area of high pollution, the first effect noted is diminished average thallus size, then a failure to produce apothecia, and finally, local extinction (Gilbert 2000).

Slow growth. Combining long life and bioaccumulation, lichens are "good summarizers of the environment" (Wetmore 1984).

Readily available. Lichens are available almost everywhere and in all seasons and permit long-term monitoring without the use of expensive equipment.

A living organism's reaction. Like the canary in the mine, they show the reaction of a living organism to pollution.

Four general approaches to lichen pollution monitoring have been used:

Distribution mapping is the plotting of the occurrence of selected species or groups of species around a suspected pollution center (Hawksworth and Rose 1970; Metzler 1980; Richardson 1992). Several possible subsets in this category range from the most simple, which involve mapping presence or absence of a single species, to a display of more complex data involving several species and multiple levels of damage.

Transplant studies. Healthy lichens can be transplanted to areas of suspected high pollution, and their health and longevity can be observed over time (Brodo 1966; Richardson 1992).

Index of Atmospheric Purity (IAP). An index can be calculated based on the number of species found at a site with adjustment for the pollution tolerance of the species found (Richardson 1992).

Heavy metal accumulation can be measured by laboratory analysis. Because lichens gather most of their required nutrients and minerals from very sparse atmospheric concentrations, they have evolved an efficient mechanism to accumulate airborne substances. Many lichen species are able to store excess accumulated substances in an inactive form, preventing them from affecting metabolic activity. These excess accumulations can then be measured with laboratory analysis providing a summary of atmospheric deposition over the life of the lichen (Purvis 2000). See Lichens and Radioactivity, page 39.

Even with organisms so well-suited to monitoring, there can be problems. A number of pollutants have been researched with lichen studies, but not all pollutants affect lichens equally. Sulphur dioxide deposition has been studied most often and seems to be the pollutant best correlated with lichen health (Metzler 1980; van Dobben and ter Braak 1999). This is convenient for monitoring purposes, since sulphur dioxide is a major component of the acid rain so detrimental to the health of our northeastern ecosystems. Identification of the lichens can also pose problems, since populations under stress often show a variable morphology.

Arctoparmelia centrifuga
Target lichen or Concentric ring lichen

light yellow-green

25 mm

Gulfside Trail at Sphinx Col, White Mountains, N.H.

This yellow-green foliose lichen has very small (1–2 mm wide), overlapping lobes, no soredia, no isidia, and a white underside. The familiar concentric pattern that is a common sight in the alpine zone is produced through radial growth, interior dieback, and interior recolonization. Although this species likes light, the best colonies are on the partially shaded north sides of rocks.

SIMILAR SPECIES: The similar *A. incurva* is much less common, and has a brown lower surface and soredia on the upper surface.

RANGE: Circumpolar arctic, and in alpine zones south to New England.

REPRODUCTION: Occasional apothecia with no soredia or isidia.

NAME: *Arctoparmelia* (L) = An arctic genus related to *Parmelia*, *centrifugus* (L) = developing from the center outwards; that is, a lichen of the north with a growth pattern developing from the center outwards.

Xanthoria elegans   Elegant sunburst lichen

Rock wall at the Greenleaf Hut, Mount Lafayatte, White Mountains, N.H.

orange

Notice the very tightly attached, convex lobes of this lichen, which is foliose, but occasionally appears as tightly attached to its substrate as some crustose species. This bright lichen seems to crave an alkaline environment and it can be found on rocks enriched by bird droppings. It appears that it is also happy with the basic chemistry of mortar, and in our northeastern alpine zones, it can be found most commonly on rock hut walls.

SIMILAR SPECIES: *Xanthoria sorediata* is similar but its cortex breaks down into patches of granular soredia, and apothecia are rare.

RANGE: Circumglobal arctic to temperate.

REPRODUCTION: Orange apothecia are common.

NAME: *Xanthoria* (G) = yellow, *elegans* (L) = elegant; that is, an elegant yellow (orange?) lichen. Some lichens in this genus are yellow-orange.

Melanelia hepatizon Rimmed camouflage lichen

Gulfside Trail on the Mount Washington cone, White Mountains, N.H.

A dark brown, foliose lichen with narrow (~1 mm), shiny lobes with thickened and raised edges, and occasional white spots (pseudocyphellae) on the lobe margins and on margins of apothecia, which are only occasionally present. The underside is black in the center, sometimes shading to dark brown near the perimeter, with sparse rhizines concentrated near the edge. Its preferred habitat is non-calcareous rock at high elevation.

SIMILAR SPECIES: *Melanelia commixta* is a near twin but the underside is pale to brown (not black). ***Melanelia stygia*** (next page) is also similar, but its lobes are convex and without raised edges, it is more black, and the pseudocyphellae are on the lobe surfaces.

RANGE: Circumglobal arctic with disjunct populations at the Great Lakes and the alpine zones of the Northeast.

REPRODUCTION: Brown to red-brown apothecia up to 5 mm across have smooth to bumpy edges (very bumpy in the detail shot above) and are concentrated near the center of the thallus.

NAME: *Melanelia* (G) = black, *hepaticus* (G) = liver-colored; that is, a very dark lichen with liver-colored thallus or apothecia.

Melanelia stygia   Alpine camouflage lichen

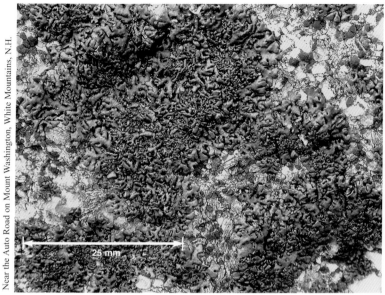

Near the Auto Road on Mount Washington, White Mountains, N.H.

dark brown to black

This foliose thallus is composed of narrow (up to 2 mm wide), convex lobes with occasional white spots (pseudocyphellae) on the lobe surface, and no soredia or isidia. The fairly common red-brown apothecia are up to 8 mm across with very bumpy rims. This is a common high-elevation lichen on exposed non-calcareous rock in both the East and the West.

SIMILAR SPECIES: See ***M. hepatizon***, previous page.

RANGE: Circumpolar arctic and boreal with alpine populations south to Washington, Montana, and North Carolina.

REPRODUCTION: Red-brown apothecia up to 8 mm across with bumpy rims are common.

NAME: *Melanelia* (G) = black, *stygos* (G) = the dark lower world; that is, a very dark lichen.

Vulpicida pinastri Powdered sunshine lichen

yellow

Gulfside Trail just north of Mount Jefferson, White Mountains, N.H.

This very small but startlingly bright foliose lichen grows on krummholz bark at medium to high elevations. Its lobes vary from 0.5 to 5 mm wide, with a yellow interior and ruffled edges covered with masses of bright yellow soredia. The whole thallus is usually less than 3 cm across but is easy to see in the otherwise dark krummholz. While its preferred substrate is conifer bark, it can be found on other species, and on dead wood. This species contains vulpinic acid, a poison that has been used to kill foxes and wolves (Brodo et al. 2001). Vulpinic acid is the pigment that produces the bright yellow color.

SIMILAR SPECIES: *Allocetraria oakesiana* is another small, conifer-bark foliose lichen with a yellowish green thallus, crisped edges, and marginal soredia, but the soredia are whitish yellow as opposed to bright yellow, and the interior is white.

RANGE: Widespread boreal to arctic.

REPRODUCTION: Dense yellow soredia are produced, and apothecia are rare.

NAME: *Vulpi* (L) = fox, *cide* (L) = kill, *pinastri* (L) = small pine; that is, referring to its fox-killing chemical content (vulpinic acid) and conifer substrate.

Hypogymnia physodes   Hooded tube lichen

light gray

East slope of Mount Washington, White Mountains, N.H.

25 mm

The thallus forms small rosettes or clumps, 2 to 6 cm across with hollow, inflated-appearing lobes, gray on top and brown to black on bottom, with no rhizines. The lobe ends break open exposing masses of soredia inside the turned-back tips, as shown in the detail photograph.

This very common, low-elevation, conifer-bark species is not usually considered to be an alpine zone lichen, but it frequently occurs on krummholz bark at least to 4,500 feet. While its preferred substrate is conifer bark, it also might be found occasionally on deciduous trees, soil, or with moss.

SIMILAR SPECIES: This is the only lichen likely to be encountered with inflated-appearing lobes and soredia inside the lobe tips.

RANGE: Northern arctic, boreal, and temperate zones, and south in the East in the high-elevation Appalachians.

REPRODUCTION: Reproduction is by soredia, and apothecia are rare.

NAME: *Hypogymnia* (G) = naked beneath, *physodes* (G) = in the form of an air sac or bladder; that is, no rhizines and puffed-up lobes.

Parmelia saxatilis — Salted shield lichen

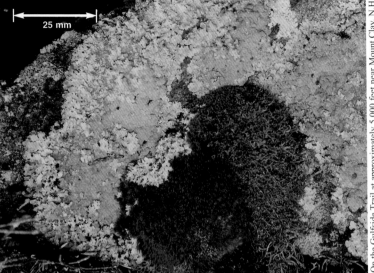

On the Gulfside Trail at approximately 5,000 feet near Mount Clay, N.H.

This common gray foliose lichen has lobes 2 to 6 mm wide containing a network of whitish ridges (pseudocyphellae) and depressions. The central portions of the thallus are covered densely with brown-tipped isidia, those micro-hot dog–shaped reproductive structures shown in the close-up photo on the right. The underside is black in the center shading to brown at the edges, with black rhizines that may be unbranched or sparsely branched, but do not have copious side branches (see below). It is listed here as a krummholz lichen, though you might encounter this generalist on soil, among moss, or on rock as its name implies.

SIMILAR SPECIES: *Parmelia saxatilis* looks like the very common but not usually alpine *P. sulcata,* which has soredia on the surface instead of isidia. *Parmelia squarrosa*, also isidiate, has rhizines with many side branches (bottle brush look) and is not alpine.

RANGE: Circumpolar arctic and boreal, south in the East following the Appalachians to North Carolina, and in the West to Arizona.

REPRODUCTION: Isidia are dense, and apothecia are uncommon.

NAME: *Parmelia* (L) = like a small round shield, *saxatilis* (L) = found with rocks; that is, a lichen of the genus *Parmelia* growing on rocks.

Parmeliopsis hyperopta Gray starburst lichen

light gray

Greenleaf Trail, Franconia Ridge, White Mountains, N.H.

25 mm

This lichen forms small rosettes on krummholz bark that can be up to 6 cm across, but are usually smaller. The lobes are small (less than 1 mm wide) and lie flat against the bark. Abundant soredia clump together in large, rough-textured mounds that are generally in the interior of the rosette, and are not found on the tips of the outermost lobes.

SIMILAR SPECIES: ***Parmeliopsis hyperopta*** and ***P. ambigua*** are essentially twin species differing only in color and chemistry.

RANGE: Circumpolar in the boreal forest, and south in Eastern North America to the Appalachian Mountains of West Virginia.

REPRODUCTION: There are large clumps of soredia, and apothecia are uncommon.

NAME: *Parmeliopsis* (L) = looks like *Parmelia*, *hyper* (G) = above. Prior to 1861, this genus was considered part of the genus *Parmelia*. Derivation of the species name is unclear.

Parmeliopsis ambigua Green starburst lichen

yellow-green

25 mm

Greenleaf Trail, Franconia Ridge, White Mountains, N.H.

This species looks like the *P. hyperopta* shown on the previous page, except that it is yellow-green instead of gray. These two closely related species often will share space on a krummholz twig. The clumped mounds of soredia in the detail photo are typical of both *P. hyperopta* and *P. ambigua*.

SIMILAR SPECIES: *Parmeliopsis hyperopta* and *P. ambigua* are essentially twin species differing only in color and chemistry. *Parmeliopsis capitata* is similar to *P. ambigua* in color and style, but has round clumps of soredia on the tips of the outermost lobes.

RANGE: Circumpolar arctic and boreal, and south in the Northeast at least as far as the White Mountains of New Hampshire.

REPRODUCTION: There are large clumps of soredia, and apothecia are uncommon.

NAME: *Parmeliopsis* (L) = looks like *Parmelia*, *ambiguus* (L) = uncertain. Prior to 1861, this genus was considered part of the genus *Parmelia*. Derivation of the species name is ambiguous.

Tuckermannopsis sepincola
Chestnut wrinkle-lichen

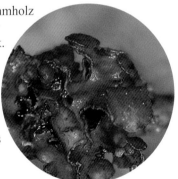

This small foliose lichen of the krummholz zone has rounded lobes that are very upright, giving it a ruffled, busy look. The apothecia rising from the upper side of the lobes are frequently so profuse that they block the view of the thallus lobes. When dry, the color is chestnut brown (above photo), but when wet the thallus gets slightly yellow.

SIMILAR SPECIES: The similar *T. fendleri* turns green when wet, and is not found in alpine or subalpine habitats.

RANGE: Arctic Canada to northeastern United States.

REPRODUCTION: Profuse apothecia, but no soredia or isidia.

NAME: *Tuckermannopsis* = for the botanist Edward Tuckerman, 1817–1886, *sepi* (L) = hedge, *cola* (L) = dwell; that is, living on hedge twigs.

Lichens and Harsh Environments

Lichens do well where climatic conditions are at their worst. In Antarctica, there are only two species of flowering plants and over two hundred species of lichen. In the Canadian Arctic, lichens take over as the dominant life form north of the treeline. And in the northeastern United States, lichens are just hitting their stride above treeline where most plants are struggling for survival. Lichens are found on all continents and can survive in deserts as well as in the cold north. What special adaptations do they have that allow them to thrive in such hostile environments? Over the four hundred-plus million years that they have been on this planet, they have evolved some very special means of coping.

Cryptobiosis. Lichens' most important adaptation to life in the tough lane is cryptobiosis, or the ability to shut down almost completely all metabolic activity through desiccation (Wessels 2001). The lichen motto is "when the going gets tough, the tough shut down." When a cryptobiotic lichen gets a bit of light and a few drops of moisture, or just some humidity in the air, it can spring into action and begin to photosynthesize. Cryptobiosis sustains them through the icy cold of the polar regions, the trials of wind-swept alpine summits, and the dry heat of deserts. Underlying this impressive ability to tolerate drought, heat, and cold stress is a cellular protective and repair mechanism that may not be well understood, but which works very well indeed (Honegger 1998).

Cold-weather photosynthesis. Some lichens can photosynthesize at temperatures as low as –20C (–4F) (Purvis 2000).

Sunscreen. Usnic acid, a common chemical in lichens, can act as a sunscreen, protecting their delicate algae or cyanobacterial partners from damage. Studies of lichens in Antarctica show that the lichens increase and decrease production of their sunscreen chemical as the day length changes (Purvis 2000).

Unique nutrient acquisition. Lichens obtain all the nutrients they need from air, atmospheric deposition in the form of rain, snow, and humidity, or from water washing along the substrate. Many lichens, particularly those that colonize nutrient-poor environments, also have nitrogen-fixing cyanobacteria as either their primary or secondary photobiont, giving them freedom to colonize sites lacking this important nutrient (Purvis 2000).

Humid is wet enough. Desert lichens can photosynthesize from damp air alone—they don't need rain. When desert air cools at night, dew forms on the lichen, so that by morning they are moistened and ready to photosynthesize. As soon as they dry up again, they slip into a cryptobiotic state and wait until the next chance to make some food. Alpine zone lichens also benefit from high humidity. The frequent fog in our northeastern alpine zones permits lichens to be active on many days when there isn't any precipitation at all.

Microscopic anchors. Fungal hyphae are so fine that they can grow between rock crystals, allowing lichens to anchor on seemingly smooth rock surfaces where nothing else can hold on.

The result of these adaptations is an organism that gets what it needs from the air, can grow on smooth rock surfaces, tolerates wide extremes of temperature and exposure, and can colonize places where no other living thing could exist. With these attributes, lichens are the first stage in colonizing abiotic environments such as glacially exposed rock, rockslide areas, or recent lava flows. This colonization by lichens frequently initiates a series of stages that can eventually lead to the formation of soil capable of supporting higher plant life (Wessels 2001).

Lichens and Radioactivity

Lichens are not an important source of direct nutrition for humans, but a great deal of lichen is consumed indirectly. Caribou are a very important food source for the Inuit people throughout the North American arctic regions, and the caribou's domesticated siblings, the reindeer, are equally important as food to the Lapp people of northern Scandinavia. In fact, lichens are the most important food for caribou and reindeer, accounting for over 60 percent of their winter diet (see Lichens as Food, p. 2).

This ecological distinction has had seriously negative consequences since World War II, because lichens have one important trait that few other living things possess. They are very tolerant of radioactivity and can survive far more radiation than vascular plants (Purvis 2000). During the 1950s and 1960s, airborne nuclear testing in the United States, the Soviet Union, and China introduced large amounts of radioactive material into the atmosphere. Radioactivity accumulated in lichens . . . caribou and reindeer ate the lichens . . . aboriginal people of the arctic ate the caribou and reindeer . . . and people thus accumulated significant amounts of radioactivity (Richardson 1975). The cessation of above-ground nuclear testing ended much of the problem until the disaster at Chernobyl in 1986. Following the Chernobyl accident, lichens in some parts of the arctic showed a 165-fold increase in radioactivity (Richardson 1992). As a result of Chernobyl, over 500 tons of reindeer carcasses were destroyed as toxic waste. The lives of the Lapp people of northern Europe were changed forever as they were forced to move from a subsistence economy to living on government handouts (Purvis 2000).

Lasallia papulosa — Toadskin lichen

gray to dark brown

Gulfside Trail at Sphinx Col, White Mountains, N.H.

This thin, fragile, gray to dark brown thallus can grow as large as 15 cm in diameter but is more commonly from 4 to 8 cm. The best identifying features are the prominent warts on the upper surface that have corresponding depressions on the lower surface, which is brown to tan without rhizines. This is a very common lichen on exposed non-calcareous rocks and ledges.

SIMILAR SPECIES: Upper surface bumps with corresponding lower surface depressions separate this from other umbilicate lichens.

RANGE: Primarily in northeastern North America with scattered disjunct populations east of Hudson's Bay, near Great Slave Lake, in coastal British Columbia, and in the southern Rockies.

REPRODUCTION: Apothecia up to 2.5 mm diameter are common, black, and with a smooth disk.

NAME: *Lasallia* = an unknown reference, probably to a person named Lasalle, *papulosa* (L) = many pimples; that is, a pimple-covered lichen.

SPECIAL TOPICS: Lichens as Food, page 2.

Umbilicaria proboscidea — Netted rock tripe

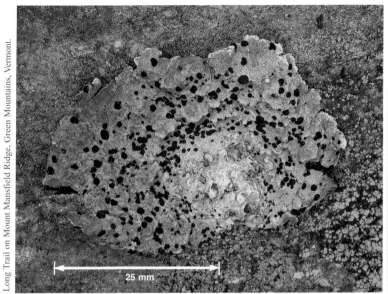

Long Trail on Mount Mansfield Ridge, Green Mountains, Vermont.

gray to brown

This variable lichen ranges from 1 to 10 cm across, is light gray to brown to almost black, and can be found either with many apothecia as above, or frequently with none. The underside is tan to gray, with or without rhizines. The best identification characteristic is the network of prominent ridges in the white crystal-covered center of the upper surface of the thallus. As with most umbilicates, this lichen has a single central attachment point. Its preferred substrate is exposed non-calcareous rock.

SIMILAR SPECIES: The crystal deposits and the network of ridges fading toward the perimeter make this a straightforward identification.
RANGE: Circumpolar arctic, and in alpine zones south to northeast United States in the East, and Washington State in the West.
REPRODUCTION: When present, apothecia are black, round, 0.5 to 1.5 mm across with clustered interior ridges.
NAME: *Umbilicaria* (L) = possessing a navel, *proboscidea* (L) = a terminal projection like an elephant's trunk; That is, the area over the central attachment looks like a navel and the wrinkles on the surface resemble the surface of an elephant's trunk.
SPECIAL TOPICS: Lichens as Food, page 2.

Umbilicaria torrefacta   Punctured rock tripe

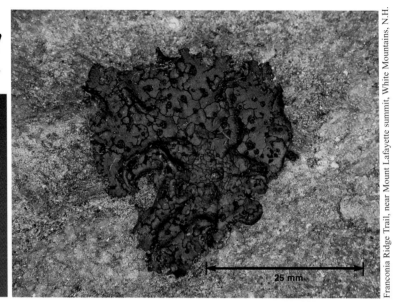

This species can grow up to 6 cm in diameter but is usually much smaller in the alpine zone. The thallus is gray to dark brown to black, and is very thin with torn and perforated edges. Apothecia are common and have concentric ridges, and the thallus underside has modified rhizines that have flattened into plate-like structures, as shown in the detail photo. The preferred substrate is exposed non-calcareous rock.

SIMILAR SPECIES: ***Umbilicaria muehlenbergii*** (p. 45) is the only other rock tripe in our area with plate-like structures, but it has a smooth, shiny surface, and no marginal perforations.

RANGE: Circumpolar arctic-alpine extending south to our northeastern alpine zones in the East, and to California in the West.

REPRODUCTION: Apothecia are common, with black disks up to 2 mm across with concentric ridges and uneven margins.

NAME: *Umbilicaria* (L) = possessing a navel, *toretos* (G) = pierced, *factus* (L) = to make; that is, an umbilicate lichen with perforations.

SPECIAL TOPICS: Lichens as Food, page 2.

Umbilicaria hyperborea Blistered rock tripe

brown to black

Thallus 2 to 5 cm across with margins that are sometimes torn. The upper surface is bumpy and deeply convex with pushed-up ridges and a crowded, wormy look. The lower surface is dark brown to black, smooth, and without rhizines. Its preferred substrate is exposed non-calcareous alpine rock. See the upper right portion of the habitat photo on page 60 for another look at this species.

SIMILAR SPECIES: This is the only umbilicate in our area with the pushed-up ridges and a smooth underside lacking plates or rhizines.

RANGE: Circumpolar arctic extending south to New Mexico in the West, and to the alpine zones of the Northeast.

REPRODUCTION: Black apothecia with concentric ridges are common.

NAME: *Umbilicaria* (L) = possessing a navel, *hyperborea* (G) = of the extreme north; that is, an umbilicate lichen of the arctic.

SPECIAL TOPICS: Lichens as Food, page 2.

Umbilicaria deusta — Peppered rock tripe

brown to black

Hamlin Ridge Trail, Mount Katahdin, Baxter State Park, Maine.

25 mm

As is true for many rock tripes, the thallus is thin, brittle, and dark brown to black. This is a small species (rarely over 5 cm diameter) with an upper surface that has a rough, textured look because of isidia that vary from occasional pepper-like granules to a rough sooty mat as shown on the detail photo. It is connected to its substrate, non-calcareous rock, by a central umbilicus.

SIMILAR SPECIES: The other centrally attached umbilicates common in our alpine zones are ***U. proboscidea***, which has a network of ridges, ***U. hyperborea***, which has a crowded, wormy look, and ***Lasallia papulosa***, which has a warty surface. No other alpine rock tripe has isidia or granules on the surface.

RANGE: Circumpolar arctic and boreal, south to the alpine zones of the Northeast.

REPRODUCTION: Both isidia and apothecia are common.

NAME: *Umbilicaria* (L) = possessing a navel, *deustus* (L) = burned up; that is, a black, charred-looking umbilicate lichen.

SPECIAL TOPICS: Lichens as Food, page 2.

Umbilicaria muehlenbergii Plated rock tripe

brown to black

Knife Edge, Mount Katahdin, Baxter State Park, Maine.

25 mm

This dark brown to black rock tripe has a satiny sheen and an underside covered with attachment plates that look like modified rhizines (detail photo). Its preferred substrate is exposed non-calcareous rock.

SIMILAR SPECIES: The satin sheen should separate this species from the cracked and areolate *U. torrefacta*, the only other umbilicate in our area with a network of attachment plates. Also, when present (infrequently), *U. muehlenbergii* apothecia have radiating ridges versus concentric ridges in *U. torrefacta*.

RANGE: Canadian boreal to maple forests, south in the mountains to North Carolina and Tennessee.

REPRODUCTION: When present, apothecia are black, sunken into depressions in the thallus, and have radiating ridges.

NAME: *Umbilicaria* (L) = possessing a navel, *muehlenbergii* is for Henry Muhlenberg (born Muehlenberg), 1753–1815, a pioneer botanist and Lutheran minister in Pennsylvania.

SPECIAL TOPICS: Lichens as Food, page 2.

Lichens and Wildlife

The Special Topic on Lichen Substances (p. 21) mentions that fungi acquired the lichen habit at least 400 million years ago, giving them ample time to evolve defenses against predatory organisms. Of course, the converse is also true—for the last 400 million years, many living things have had opportunities to evolve strategies to get at those nutritious chunks of carbohydrate, or to find other uses for the highly diverse lichen species.

Caribou, reindeer (domesticated caribou), deer, and elk are major consumers of lichen on the arctic tundra and in boreal to subalpine woodland. During the winter when lichen is their primary food, each caribou consumes approximately 3 kilograms per day (Gilbert 2000; Richardson 1975). (See Lichens as Food, p. 2; Lichens and Radioactivity, p. 39).

Many birds also have developed a close relationship with lichens. Ruby-throated hummingbirds weave scraps of the gray foliose lichen *Parmelia sulcata* into their nests, presumably as camouflage. Golden plovers, sandpipers, and ptarmigans all breed on the arctic tundra and use the whiteworm lichen (***Thamnolia vermicularis***), common in the northeastern alpine zones, for nest-building (Brodo et al. 2001; Pielou 1994). The parula warbler is more flexible in its choice of material. It uses *Usnea* species (beard lichens) in the northern part of the United States where *Usnea* is common and *Ramalina,* or even Spanish moss (not a lichen), in the southern United States where *Usnea* isn't available (Richardson 1975; Sibley 2000). Studies of the correlation between breeding success of birds and availability of suitable nest-building materials indicate that the presence of an appropriate lichen flora is a potentially important factor in breeding success (Gilbert 2000). And it's not just birds that use lichens as nest material; flying squirrels and bats use horsehair lichen (*Bryoria* spp.) in building their nests (Richardson 1975).

Lichens, along with their algal partners and moss cohorts, form the basis of the tree-trunk food web and provide housing, food, and camouflage for many organisms. Herbivorous arthropods graze on lichens, carnivorous arthropods (and others) dine on the herbivores, and birds dine on both the herbivores and the carnivores. The birds then help re-establish the base of this food web by accumulating lichen soredia in their feathers as they feed, spreading the propagules to new sites (Gilbert 2000).

Crustose Lichen Identification

The world of lichens usually is divided into two large subgroups, the macrolichens and the microlichens. The shrubby, stalked, and foliose lichens that we've seen up to this point in the book are all included in the macrolichen subgroup, and most can be readily identified by a careful inspection with nothing more than a hand lens.

The crustose lichens that follow all fall within the microlichen subgroup and can be much more difficult to identify. Crustose lichen identification often requires a microscope so that features of the fruiting bodies, especially the spores, can be examined. Since crustose lichens do not have a lower cortex or skin, and their fungal filaments actually penetrate the substrates they colonize, sampling a crustose lichen requires that you collect substrate as well as lichen. In the case of rock-dwelling lichens, this requires use of a rock hammer and cold chisel in the field. This certainly can make crustose identification more complicated for a beginning lichenologist.

For those reasons, the species identification pages 48 to 63 should be considered as tentative guides not providing definitive identifications. Also, of the well over one hundred crustose lichens that might be found in our alpine zones, we have included only a few of the species that seem to lend themselves at least in part to identification with nothing more powerful than a hand lens. Some, such as the ***Ochrolechia frigida*** on the next page and the ***Rhizocarpon geographicum*** on page 54, can be reliably identified visually, but others, like the soil crust on page 50, the orange/red rock species below and on pages 55 to 57, or the individual species within the ***Lecanora subfusca* group** discussed on page 62, can be very challenging.

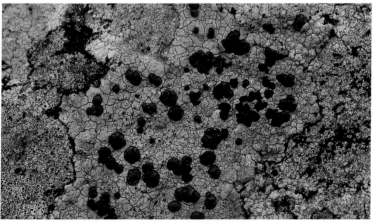

Porpidia flavocaerulescens on Franconia Ridge (see p. 55).

Ochrolechia frigida — Arctic saucer lichen

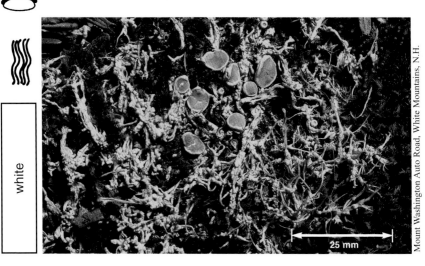

This white thallus, usually with spine-like projections, grows agressively over soil, moss, or dead vegetation. Apothecia are common (up to 5 mm across) with white margins and pink disks. Two different forms of this variable species are shown below.

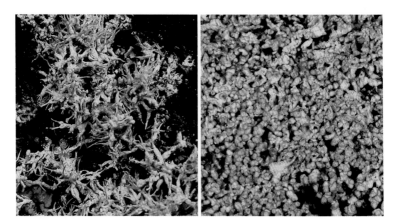

SIMILAR SPECIES: Nothing in the alpine zone is similar.

RANGE: Circumpolar arctic-alpine, south in the East to the alpine zones of the Northeast.

REPRODUCTION: Apothecia are common, soredia are occasional.

NAME: *Ochro* (G) = brownish yellow, *leche* (Spanish from *lacteus* [L]) = milk white, *frigida* (L) = cold; that is, a cold-adapted lichen with mustard-yellow apothecia on a white thallus.

Icmadophila ericetorum — Candy lichen

pale green

Gray Knob Trail, White Mountains, N.H.

With a pale green thallus and pink apothecia (1 to 3 mm across), this is a very striking lichen and an easy identification. It grows in moist, shaded areas on soil, moss, and rotted wood. Look for it under the krummholz along wet, shady trail edges. This is a striking and easily recognized lichen.

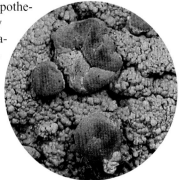

SIMILAR SPECIES: *Dibaeis baeomyces* is a similar, related soil lichen, but its primary thallus is whiter, the apothecia are round and clearly stalked, and it is unlikely in the alpine zone.

RANGE: Circumpolar arctic-alpine to the summits of New England and the Rocky Mountains.

REPRODUCTION: Pink apothecia are common with or without a pale margin. There are no soredia or isidia.

NAME: *Icmadophila* (G) = moisture loving, *ericetorum* (G) = of the heath; that is, moisture-loving and growing among heath plants.

Trapeliopsis granulosa — Mottled-disk lichen

light gray to greenish gray

East side of Mount Washington, White Mountains, N.H.

25 mm

This pale crustose thallus will grow over soil, moss, old wood, or occasionally pebbles. It is very lumpy, with pink to brown convex (sometimes hemispherical) apothecia similar in form to the granules of the thallus. In bright sunlight the color may be darkened considerably. This important colonizer of bare soil is a pioneer species after fire. In Canada, it has been introduced to burned areas to stabilize soil and reflect sunlight, allowing moisture to accumulate (Brodo et al. 2001).

SIMILAR SPECIES: *Lecidoma demissum*, another earth-stabilizing crust, is darker and less bumpy, with dark brown to black apothecia, and a sometimes lobed margin.

RANGE: Throughout North America, with the exception of the Southeast from Texas to South Carolina and south.

REPRODUCTION: Apothecia color can range from pink, to red-brown, to white, to dark gray-green, explaining its earlier specific epithet, *quadricolor* (Gilbert 2000). It frequently produces coarse granular soredia.

NAME: *Trapeliopsis* (G) = looks like *Trapelia*, *granulosa* (L) = full of small grains; that is, a fine-grained lichen that looks like species in the genus *Trapelia*.

Aspicilia cinerea — Cinder lichen

Ash gray

The light gray thallus can be either thin or thick, often thicker near the center, and thinner and almost lobate toward the perimeter, breaking up throughout into small polygons (areoles) approximately 0.5 to 2 mm across. Flat, black apothecia, usually no more than one per areole, are frequently sunken into the areoles but sometimes emerging and developing margins in older portions of the thallus. It grows on siliceous rocks, usually in full sun.

SIMILAR SPECIES: *Aspicilia verrucigera* is similar but has a thicker and more bumpy thallus which can appear lobed in older specimens.

RANGE: Common over much of North America.

REPRODUCTION: The apothecia disk is usually level with the surrounding thallus. There are no soredia or isidia.

NAME: *Aspicilia* = derivation unknown, *cinerea* (L) = ash-colored; that is, an ash-colored lichen.

Lecanora polytropa   Granite-speck rim-lichen

light yellow

Long Trail on Mount Mansfield Ridge, Green Mountains, Vermont.

Growing on rock with hyphae penetrating microscopic rock fractures, this thallus is only occasionally in view as completely as in the photograph above. Frequently, the light yellow areoles are more dispersed, leaving not much more than the domed waxy-yellow apothecia in view, as shown in the magnified detail. This is a wonderful example of a lichen thriving where nothing else could survive. This species is very common on granitic rocks in full sun.

SIMILAR SPECIES: This is a very common lichen that is relatively easy to identify with a bit of patience and a 10X lens.

RANGE: Circumpolar arctic and boreal, following high terrain south to Mexico in the West, and to North Carolina in the East.

REPRODUCTION: Apothecia are domed to flat, usually waxy-yellow with a pale margin. There are no soredia or isidia.

NAME: *Lecanora* (G) = dish-shaped with border, *polytropa* (G) = many turns; that is, apothecia that are dish-shaped with a twisted margin.

Ophioparma ventosa   Alpine bloodspot

cream to gray

Mount Clay, White Mountains, N.H.

This lichen, with its thick, lumpy, creamy-white to yellowish thallus and blood-colored apothecia with thallus-colored margin, is very common on non-calcareous rocks in full sun. It's hard to miss.

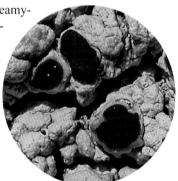

SIMILAR SPECIES: This is a very conspicuous lichen and an easy identification.

RANGE: Primarily in arctic Canada, plus a disjunct population in northeastern United States.

REPRODUCTION: The distinctive apothecia are up to 3 mm across. There are no soredia or isidia.

NAME: *Ophio*s (G) = snake, *parma* (L) = shield, *ventosa* (L) = windy; that is, a shield shaped lichen of windblown places. The snake reference is probably to the snake-shaped spores.

Rhizocarpon geographicum — Map lichen

yellow

Col between Mount Clay and Mount Jefferson, White Mountains, N.H.

25 mm

If there's a signature lichen of our northeastern alpine zones, this is it. The thallus consists of bright yellow to yellow-green areoles grouped together and surrounded by a black prothallus (margin) that outlines areole groups and creates shapes that might resemble countries on a map. It thrives in full sun on the non-calcareous rocks of the northeastern alpine zones. This is a very pollution-tolerant lichen that will recolonize quickly following reduction of a pollutant (McCune and Geiser 2000).

SIMILAR SPECIES: There are several yellow map lichens in arctic-alpine North America, but this is by far the most widespread (Brodo et al. 2001).

RANGE: Widely distributed from the high arctic to temperate zones of both North and South America, but not extending south of the northeastern alpine zones in the East.

REPRODUCTION: Small black apothecia (to 1.5 mm across) appear between areoles and are hard to distinguish from the black prothallus There are no soredia or isidia.

NAME: *Rhizocarpon* (G) = fruiting directly from the root, *geographicum* (G) = geographic, or map-like; that is, a lichen with map-like shapes and apothecia appearing to rise directly from the substrate.

Porpidia flavocaerulescens Orange boulder lichen

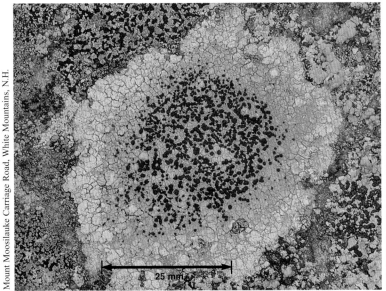

Rusty orange crustose lichens on rock are a difficult group in the alpine zone because iron in the rock can cause a number of species to take on a rusty coloration. I've include three species here and on the next two pages, but a certain amount of caution is appropriate. ***Porpidia flavocaerulescens*** is orange occasionally shading to gray, with an irregularly cracked surface and black margined, usually convex, black apothecia becoming bluish gray because of a light pruina or crystalline deposit on the disk surface.

SIMILAR SPECIES: See next two pages.

RANGE: This is a circumpolar arctic lichen with disjunct populations in the alpine zones of the Northeast.

REPRODUCTION: Black apothecia are up to 3 mm diameter with black margins. There are no soredia or isidia.

NAME: *Porpidia* (G) = like a small brooch or clasp, *flavidus* (L) = yellow, *caerulescens* (L) = becoming blue; that is, a brooch-shaped lichen with a yellowish (orange?) thallus developing blue-black apothecia.

Tremolecia atrata — Rusty-rock lichen

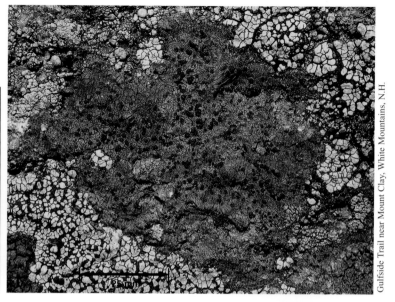

Gulfside Trail near Mount Clay, White Mountains, N.H.

This rusty red rock lichen has a fairly thin, smooth rusty red thallus only slightly cracked into areoles and surrounded by a distinctive black prothallus edging. The apothecia are very small (< 0.5 mm), black, and immersed in the thallus. Its preferred habitat is non-calcareous rock in full sun.

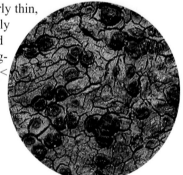

SIMILAR SPECIES: ***Porpidia flavocaerulescens*** has larger convex apothecia, a less reddish thallus, and no prothallus, while ***Lecidea lapicida*** usually has some gray in it, and no prothallus.

RANGE: Circumpolar arctic, south throughout the Rockies in the West, and with a disjunct population in the Alpine zones of the Northeast.

REPRODUCTION: The black, immersed apothecia are very small (< 0.5 mm across). There are no soredia or isidia.

NAME: *Trema* (G) = immersed, *lekos* (G) = disk, *atrata* (L) = black; that is, immersed, black, disk-shaped apothecia.

Lecidea lapicida — Gray-orange disk lichen

Gulfside Trail just south of Mount Jefferson, White Mountains, N.H.

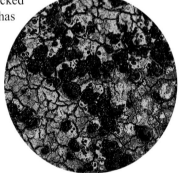

Notice how this thin and slightly cracked rusty red thallus shades to gray and has no prothallus. Depending on substrate, it is sometimes all gray. The small black apothecia (up to 1.5 mm) have distinct black margins, and can be convex, to slightly concave. Like the other rusty rock lichens, its preferred habitat is non-calcareous rock in full sun.

SIMILAR SPECIES: ***Porpidia flavocaerulescens*** has larger, convex apothecia and a more uniformly orange thallus, while ***Tremolecia atrata*** usually has no gray patches and has a distinct prothallus.

RANGE: Circumpolar arctic and boreal, ranging south in the East to the mountain tops of the Northeast, and in the West to Washington.

REPRODUCTION: The black apothecia are up to 1.5 mm across with black borders. There are no soredia or isidia.

NAME: *Lecidea* (G) = small dish, *lapicida* (L) = on rock; that is, a rock lichen with small dish-shaped apothecia.

Acarospora fuscata Brown cobblestone lichen

dark brown

25 mm

A thick, shiny, dark brown thallus consisting of areoles that are frequently lifted at the edges. Each areole may have several very small (< 1.0 mm), immersed, red-brown apothecia. The preferred substrate is granitic rock. Note in the photograph above the mix of **Rhizocarpon geographicum** and **Orphniospora moriopsis** surrounding the *Acarospora fuscata*.

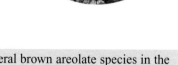

SIMILAR SPECIES: There are several brown areolate species in the genus *Acarospora*, but this is the most common (Brodo et al. 2001).

RANGE: Common throughout North America, except in southeastern United States south of North Carolina.

REPRODUCTION: Small, dark, red-brown apothecia are common. There are no soredia or isidia.

NAME: *Acar* (G) = small, *spora* (G) = spore, *fuscata* (L) = dark brown; that is, a brown lichen with tiny spores.

Protoparmelia badia — Chocolate rim-lichen

Gulfside Trail just north of Mount Washington, White Mountains, N.H.

dark brown

Thick, chocolate brown thallus with lumpy areoles. Apothecia are common, with dark disks to 3 mm wide and chocolate-colored rims. It is found primarily on non-calcareous alpine rocks. Notice the high magnification of the photo above.

SIMILAR SPECIES: No other crustose lichen has this rich chocolate color with distinctive dark brown apothecia up to 3 mm wide.

RANGE: North America has disjunct populations in the Rocky Mountains and in the mountains of the Northeast.

REPRODUCTION: Relatively large, dark brown (almost black when dry) apothecia are common. There are no soredia or isidia.

NAME: *Proto* (G) = first, *parmelia* (G) = the lichen genus *Parmelia*, *badia* (L) = chocolate-brown; that is, a chocolate-brown lichen in a genus thought to have developed as a precursor to the genus *Parmelia*.

Orphniospora moriopsis   Black-on-black lichen

Mount Washington just above the Alpine Gardens, White Mountains, N.H.

This is the proverbial black cat in the coal bin. It has a thick, areolate, black thallus with black apothecia and a black prothallus. The apothecia are small, and can be either immersed in the areoles or convex above them. Notice the **_Umbilicaria hyperborea_** in the upper right corner of the habitat photograph.

SIMILAR SPECIES: This is a distinctive combination of all black parts.

RANGE: Circumpolar arctic and alpine extending south to the alpine zones of the Northeast, and to Washington in the West.

REPRODUCTION: The very small, black apothecia are inconspicuous. There are no soredia or isidia.

NAME: *Orphnos* (G) = dark, *sporos* (G) = spore, *moriopsis* (L) = looks dead; that is, a lichen the color of death (black) with dark spores.

Lecanora symmicta — Fused rim-lichen

light gray to yellow-green

Whiteface Mountain summit, Adirondacks, N.Y.

This lichen thallus is thin and light gray to yellow-green with flat to convex yellow to yellow-brown apothecia having thin, smooth margins, at least when young, frequently growing into each other and joining—thus the "fused rim-lichen." It is a very common lichen on krummholz bark, but frequently the colors are less bright than in the photograph above, and it doesn't stand out.

SIMILAR SPECIES: The yellow to yellow-brown apothecia with smooth, pale, often disappearing margins, together with the thin yellowish white thallus are quite distinctive.

RANGE: Arctic, boreal, and temperate, throughout Canada and northeastern United States.

REPRODUCTION: Apothecia are common, and the thallus is sometimes sorediate. No isidia are produced.

NAME: *Lecanora* (G) = dish-shaped, *synnictus* (G) = commingled; that is, fusing apothecia.

Lecanora subfusca group   Rim-lichens

light gray to white

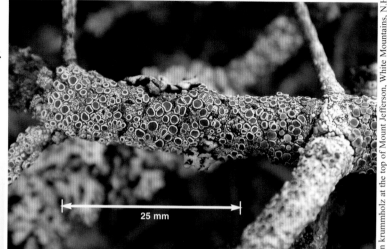

On krummholz at the top of Mount Jefferson, White Mountains, N.H.

25 mm

The ***Lecanora subfusca*** group includes at least 4 species in the alpine and subalpine zones of the Northeast that are virtually impossible to separate without a detailed microscopic analysis of the apothecial structure and the spores, so we treat them together here. The pale gray to white thallus occasionally has a black prothallus. The apothecia have a rim the same color as the thallus and a disk that is usually red-brown but occasionally black. The preferred substrate is krummholz bark.

SIMILAR SPECIES: The four species of the **L. subfusca** group known from the subalpine zone are *L. circumborealis*, *L. hybocarpa*, *L. pulicaris*, and *L. wisconsinensis*. The species most likely to be found in the alpine zone is *L. circumborealis*.

RANGE: *L. circumborealis* ranges from the arctic south to northeastern United States, *L. hybocarpa* is in eastern United States, *L. pulicaris* and *L. wisconsinensis* are boreal forest species (Brodo 1984).

REPRODUCTION: There are many very small (< 1.2 mm) orange to red-brown to black disked apothecia with thallus-colored rims. Soredia are sometimes produced, but there are no isidia.

NAME: *Lecanora* (G) = dish-shaped, *subfusca* (L) = incompletely dark or dark brown; that is, referring to the color of the apothecia, or indicating a relationship to *Lecanora fusca*.

Mycoblastus sanguinarius Bloody heart lichen

East slope of Mount Washington, White Mountains, N.H.

light gray

25 mm

This fun lichen is common on conifers of the boreal forest as well as on krummholz in the alpine zone itself. It consists of a thick, lumpy, gray-white to occasionally slightly green thallus (greener when wet), with many strongly convex, borderless, black apothecia about 2 to 3 mm diameter. If you break apart one of the apothecia (use a knife and your hand lens), you will find a blood-red base as shown in the detail photograph on the right.

SIMILAR SPECIES: There are other crustose lichens in the boreal forest with black convex apothecia, but this is the only one with a hidden red base.

RANGE: Circumpolar arctic and boreal, south to North Carolina in the East, and to Oregon in the West.

REPRODUCTION: Unusual black apothecia with a blood-red base. Soredia are sometimes produced, but there are no isidia.

NAME: *Myco* (G) = fungal, *blastus* (G) = sprout, *sanguinarius* (L) = blood-red; that is, reference to the blood-red tissue in the apothecial base.

Glossary

apothecium (pl. apothecia): A spore-producing structure that looks like a baby quiche or occasionally a beret.

areole: In crustose lichens, a small polygonal or rounded patch of thallus frequently with raised edges. Areoles can be scattered and separate (dispersed) or simply separated by deep cracks. In stalked lichens, it refers to a raised piece of cortex material on the surface of a podetium or cup.

ascoma (pl. ascomata): The spore-producing part of fungus in the Ascomycota phylum of fungi, the phylum most likely to be lichenized. The most commonly seen ascomata on alpine lichens are called apothecia (see apothecium). Perithecia are ascomata that open to the surface through a tiny hole and are therefore much less conspicuous (see "perithecium").

ascomycota: A phylum of fungi that produce ascomata as their reproductive structures. This group includes disk- or cup-fungi, and approximately half of this group form lichens.

axils: The joint between two branches. Frequently described as "open" (with a hole to a hollow interior) or "closed."

basidiomycota: A phylum of fungi that produce basidiomata as their reproductive structures. This group includes mushrooms and bracket fungi. Very few members of this group form lichens.

basidioma (pl. basidiomata): Puffball or mushroom-like reproductive structure formed by the basidiomycota group of fungi.

calcareous: Refers to rocks (or soil) containing calcium carbonate (chalk), such as limestone or marble. This high pH substrate is very uncommon in the alpine zones of the Northeast.

chemotype: Lichen populations of the same species with different chemical profiles.

cortex: A layer of tightly packed fungal material forming the upper and/or lower surface of a lichen thallus.

crustose: A lichen with no lower cortex, and so tightly adhered to its substrate that it cannot be collected without removing substrate.

cryptobiotic: The ability to shut down almost completely all metabolic activity.

disjunct: Disjunct populations of a species are separated by significant distances, that is, they are not contiguous.

foliose: A leafy growth form with a distinct upper and lower surface and a lobed shape at the perimeter.

fruticose: A basically three-dimensional growth form that includes bushy, shrubby, hanging, tufted, or stalked shapes.

isidia: Corticate outgrowths of the thallus that can be cylindrical (hot-dog–shaped), granular, or scaly. These tiny asexual reproductive packages contain both photobiont and mycobiont partners.

krummholz: A German word meaning "crooked wood." It refers both to the vegetation zone above the spruce-fir zone and to the woody vegetation there and in the alpine zone that exhibits a largely horizontal growth form.

medulla: The interior of a lichen thallus.

mycobiont: The fungal part of the lichen.

non-calcareous: Refers to rock such as granite or gneiss containing no calcium carbonate and therefore having little or no buffering capacity. This low pH substrate is the norm in our alpine zones.

***para*-phenylenediamine (P or PD)**: A staining, potentially carcinogenic spot test reagent that turns red, orange, or yellow when applied to a lichen containing certain substances.

perithecium: A vase- or flask-shaped ascoma with a small pore-like opening at the top, usually at least partially burried in the thallus, appearing like a black dot or round bump on the thallus surface.

photobiont: The photosynthetic part of the lichen.

phyllocladia: Lobule-like growths on *Stereocaulon* stalks.

podetium (pl. podetia): A usually vertical stalk. It can be highly branched and tangled as in the reindeer lichens, or more solitary as in many of the stalked lichens in the *Cladonia* genus.

potassium hydroxide: A spot test reagent abbreviated K or KOH.

prothallus: In crustose lichens, a purely fungal part of the lichen extending beyond the thallus perimeter and between areoles.

pruina: A frosty crystalline deposit on the surface of a thallus or apothecium.

pseudocyphellae: An opening in the cortex allowing the color of the medulla to show through. They frequently appear as white dots or lines on the lichen surface.

rhizines: Short filaments on the lower surface of many foliose and umbilicate lichens providing attachment to a substrate.

siliceous: Containing mainly silicon-type materials; refers to the non-calcareous (low pH) rocks typical of our Northeastern alpine zones.

sodium hypochlorite: Bleach. A spot test reagent abbreviated C.

soralium (pl. soralia): A patch or area containing soredia.

soredia: Asexual reproductive packages on the surface of a lichen thallus containing both the mycobiont and the photobiont. They can vary from very fine powder to granules.

squamules: Scale-like lobes attached by an edge. In the genus *Cladonia*, they frequently give rise to a fruticose growth form with podetia.

thallus: The main vegetative body of a lichen.

tomentose: With fuzzy, often matted hairs (tomentum).

umbilicate: A lichen growth form characterized by a central holdfast (umbilicus) and a shape resembling a potato chip. Umbilicate lichens grow on rock and are a subset of foliose since they have a distinct upper and lower cortex.

References

Borror, D. J. 1988. *Dictionary of word roots and combining forms.* Mayfield Publishing Company, Mountain View, Calif.

Brodo, I. M. 1966. Lichen growth and cities: A study on Long Island, New York. *The Bryologist* 69:426–449.

———. 1984. The North American species of the Lecanora subfusca group. In H. Hertel and F. Oberwinkler, eds. Beitrage zur lichenologie. Festscrift J. Poelt. Beiheft zur *Nova Hedwigia* 79. J. Cramer, Vaduz, pp. 63–185.

Brodo, I. M., S. D. Sharnoff, and S. Sharnoff. 2001. *Lichens of North America.* Yale University Press, New Haven, Conn.

Brown, R. W. 1956. *Composition of scientific words.* Published by the author, Baltimore, Md.

Burt, P. 2000. *Barrenland beauties.* Up Here Publishing Ltd., Yellowknife, Northwest Territories, Canada.

Campbell, N. A., J. B. Reece, and L. G. Mitchell. 1999. *Biology.* Addison Wesley Longman, Inc., Menlo Park, Calif.

Darwin, Erasmus, 1790. *The botanic garden.* J. Johnson, London. From a 1978 reprint by Garland Publishing, N.Y.

Esslinger, T. L. 1997. A cumulative checklist for the lichen-forming, lichenicolous and allied fungi of the continental United States and Canada. North Dakota State University: http://www.ndsu.nodak.edu/instruct/esslinge/chcklst/chcklst7.htm (First Posted 1 December 1997, Most Recent Update 2 March 2004), Fargo, North Dakota.

Ferry, B. W., M. S. Baddeley, and D. L. Hawksworth, eds. 1973. *Air pollution and lichens.* The Athlone Press, London.

Feige, G. B. 1998. *Etymologie der wissenschaftlichen gattungsnamen der flechten.* Botanisches Institut und Botanisher Garten der Universitat Essen - Gesamthochschule, Essen, Germany.

Fahselt, D. 1996. Individuals, populations and population ecology. In T. Nash III, ed. *Lichen Biology.* Cambridge University Press, Cambridge, pp. 181–198.

Gilbert, O. 2000. *Lichens.* Harper Collins, London, England.

Hale, M. E. 1979. *How to know the lichens.* WCB McGraw-Hill, Boston, Mass.

Hawksworth, D. L., and F. Rose. 1970. Qualitative scale for estimating sulphur dioxide air pollution in England and Wales using epiphytic lichens. *Nature* 227:145–148.

Henderson, A. 2000. Literature on air pollution and lichens XLIX. *Lichenologist* 32:89-102.

Hinds, P. L., and J. W. Hinds. 1992. *Simplified field key to Maine macrolichens.* Published by the authors, Maine.

Honegger, R. 1998. The lichen symbiosis—What is so spectacular about it? *Lichenologist* 30:1096–1135.

Jaeger, E. C. 1944. *A source-book of biological names and terms.* Charles C. Thomas, Publisher, Springfield, Illinois.

Kendrick, B. 2000. *The fifth kingdom*. Focus Publishing, Newburyport, Mass.

Marchand, P. 1987. *North woods*. Appalachian Mountain Club, Boston, Mass.

Marles, R. J., C. Clavelle, L. Monteleone, N. Tays, and D. Burns. 2000. *Aboriginal plant use in Canada's northwest boreal forest*. UBC Press, Vancouver and Toronto, Canada.

McCune, B., and L. Geiser. 2000. *Macrolichens of the Pacific Northwest*. Oregon State University Press, Corvallis, Wash.

Merrill, G. K. 1915. Notes on the primitive uses of lichens. *Bulletin Josselyn Botanical Society* 5:18–25.

Metzler, K. J. 1980. *Lichens and air pollution: A study in Connecticut*. State Geological and Natural History Survey of Connecticut, Hartford.

Nylander, W. 1896. *Les lichens des environs de Paris*. Typographie P. Schmidt, Paris.

Pielou, E. C. 1994. *A naturalist's guide to the arctic*. The University of Chicago Press, Chicago and London.

Purvis, W. 2000. *Lichens*. The Natural History Museum, London.

Richardson, D. H. S. 1975. *The vanishing lichens, their history, biology and importance*. David & Charles, London.

———. 1992. *Pollution monitoring with lichens*. The Richmond Publishing Co., Ltd., Slough, England.

Sibley, D. A. 2000. *The Sibley guide to birds*. Alfred A. Knopf, New York.

Simpson, D. P. 1968. *Cassell's Latin dictionary*. Macmillan Company, New York.

Stearn, W. T. 1992. *Botanical Latin*. Timber Press, Portland, Oregon.

Thomson, J. W. 1984. *American arctic lichens*: vol. 1. *The macrolichens*. Columbia University Press, New York.

———. 1997. *American arctic lichens*: vol. 2. *The microlichens*. University of Wisconsin Press, Madison, Wisconsin.

Uppsala University. 2002. Erik Acharius (1757–1819) "the father of lichenology." Uppsala University, Evolution Museum: www.evol-musem.uu. Accessed 11 February 2002.

van Dobben, H. F., and C. J. F. ter Braak. 1999. Ranking of epiphytic lichen sensitivity to air pollution using survey data: A comparison of indicator scales. *Lichenologist* 31(1):27–39.

Wessels, T. 2001. *The granite landscape: A natural history of America's mountain domes, from Acadia to Yosemite*. The Countryman Press, Woodstock, Vt.

Wetmore, C. M. 1984. Lichens and air quality in Acadia National Park. National Park Service. Contract CX 0001-2-0034.

Wilson, B. 2002. Personal conversation with Dr. Beverly Wilson, family physician in Yellowknife, Canada, re: Inuits and cancer rates.

Species Index

(species and page numbers in **bold** type indicate description pages)

Acarospora fuscata, **58**
Allocetraria oakesiana, 32
alpine camouflage lichen, 31
alpine bloodspot, 53
arctic saucer lichen, 48
Arctoparmelia centrifuga, **28**
Arctoparmelia incurva, 28
Aspicilia cinerea, **51**
Aspicilia esculenta, 2
Aspicilia verrucigera, 51

ball lichen, 7
black-footed reindeer lichen, 10, 13
black-on-black lichen, 60
blistered rock tripe, 43
bloody heart lichen, 63
boreal oakmoss lichen, 8
brown cobblestone lichen, 58

candy lichen, 49
Cetraria islandica, 5, 6
Cetrariella laevigata, **5, 6**
Cetrariella delisei, **6**
chestnut wrinkle-lichen, 37
chocolate rim-lichen, 59
cinder lichen, 51
Cladonia amaurocraea, 16, **17**
Cladonia arbuscula, 13, **15**, 16
Cladonia borealis, 24
Cladonia coccifera, 23, **24**
Cladonia deformis, 22
Cladonia gracilis, 17
Cladonia mitis, 15
Cladonia pleurota, 22, **23**
Cladonia rangiferina, **10**, 13, 15
Cladonia squamosa, 21, **25**
Cladonia stellaris, 13, **14**
Cladonia strepsilis, **7**
Cladonia sulphurina, **22**
Cladonia stygia, 10, **13**, 15

Cladonia uncialis, 13, **16**, 17
concentric ring lichen, 28
cornucopia lichen, 24
curled snow lichen, 4

Dibaeis baeomyces, 49
dragon cladonia, 25

earth foam lichen, 11
elegant sunburst lichen, 29
Evernia mesomorpha, **8**, 9

finger-scale foam lichen, 20
Flavocetraria cucullata, 3, **4**,
Flavocetraria nivalis, 3, **4**
fragile coral lichen, 18
fused rim-lichen, 61

granite-speck rim-lichen, 52
gray starburst lichen, 35
gray-orange disk lichen, 57
greater sulphur-cup, 22
green starburst lichen, 36

hooded tube lichen, 33
Hypogymnia physodes, **33**
horsehair lichen, 2

Iceland lichen, 2, 5, 6, 9
Icmadophila ericetorum, **49**

Lasallia papulosa, **40**, 44
Lecanora circumborealis, 62
Lecanora subfusca **group, 47, 62**
Lecanora esculenta, 2
Lecanora hybocarpa, 62
Lecanora polytropa, **52**
Lecanora pulicaris, 62
Lecanora symmicta, **61**
Lecanora wisconsinensis, 62
Lecidea lapicida, 56, **57**

Lecidoma demissum, 50
Lobaria pulmonaria, 9
lungwort, 9

map lichen, 54
Melanelia commixta, 30
Melanelia hepatizon, **30**, 31
Melanelia stygia, 30, **31**
mottled-disk lichen, 50
Mycoblastus sanguinarius, **63**

netted rock tripe, 41

Ochrolechia frigida, 47, **48**
Ophioparma ventosa, **53**
orange boulder lichen, 55
Orphniospora moriopsis, **60**

Parmelia saxatilis, **34**
Parmelia squarrosa, 34
Parmelia sulcata, 34, 46
Parmeliopsis ambigua, 35, **36**
Parmeliopsis capitata, 36
Parmeliopsis hyperopta, 35, **36**
peppered rock tripe, 44
plated rock tripe, 45
Porpidia flavocaerulescens, 47, **55**, 56, 57
powdered sunshine lichen, 32
Protoparmelia badia, **59**
punctured rock tripe, 42

quill lichen, 17

Ramalina spp., 8, 46
red-fruited pixie-cup, 23
reindeer lichen, 13, **15**
Rhizocarpon geographicum, 47, **54**
rim-lichens, 59, 61, 62
rimmed camouflage lichen, 30

Roccella tinctoria, 9
rusty-rock lichen, 56

salted shield lichen, 34
seafoam lichen, 19
snow-bed Iceland lichen, 6
snow lichen, 3
Sphaerophorus fragilis, **18**
star-tipped reindeer lichen, 14
Stereocaulon alpinum, **11**, 19
Stereocaulon dactylophyllum, 19, **20**
Stereocaulon glaucescens, **19**, 20
Stereocaulon saxatile, 19
striped Iceland lichen, 5

target lichen, 28
Thamnolia vermicularis, **12**, 46
thorn cladonia, 16
toadskin lichen, 40
Trapeliopsis granulosa, **50**
Tremolecia atrata, **56**, 57
Tuckermannopsis fendleri, 37
Tuckermannopsis sepincola, **37**

Umbilicaria deusta, **44**
Umbilicaria hyperborea, **43**, 44, 60
Umbilicaria muehlenbergii, 42, **45**
Umbilicaria proboscidea, 41, **44**
Umbilicaria torrefacta, **42**, 45
Usnea hirta, 9

Vulpicida pinastri, **32**

whiteworm lichen, 12, 46

Xanthoria elegans, **29**
Xanthoria sorediata, 29

The author, lichen hunting in Maine

RALPH POPE is an adjunct professor of New England Flora at Antioch New England Graduate School in Keene, N.H., where he also earned his M.S. in Environmental Studies.